A BOOK of MY OWN

This book is owned by:

This is a gift from your friends at
Goodwill Industries of the Columbia Willamette.

GOODWILL
INDUSTRIES
OF THE COLUMBIA WILLAMETTE

Our business is changing lives.™

More praise for

THE ELEPHANT'S SECRET SENSE

"This fascinating book reads like a fast-paced detective story of scientific discovery and adventure set in contemporary Africa, populated by a rich cast, both animal and human, detailing the joys and frustrations of a remarkable young scientist who has gained new insight into the elephants' way of life. In places very funny, although the hard facts and tragedy of life in Africa are not ignored. By the end she takes her rightful place among the leading biographers of the African elephant."

> —Iain Douglas-Hamilton, author of *Among the Elephants,* co-author of *Battle for the Elephants*, and founder of Save the Elephants

"A tightly-woven story of human warmth, strange cultures, fabulous wildlife, and scientific discovery. Only the best science and nature writing draws you into a whole different world, and this remarkable book does it right from page one—with grace, humility, and all the exotic splendor of Africa."

> —Carl Safina, author of *Song for the Blue Ocean* and *Voyage of the Turtle*

"A wonderful book about working with wonderful animals. Dr. O'Connell has opened up for us the world of elephant communication. It is one of the many interesting aspects of these giants that we have just learned also (like some of our great ape relatives) have a sense of self. If you like elephants, you'll love *The Elephant's Secret Sense*."

> —Paul R. Ehrlich, Ph.D., president, Center for Conservation Biology, Stanford University, and author of *One with Nineveh*

THE
ELEPHANT'S
SECRET SENSE

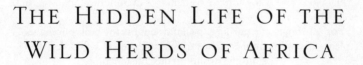

THE HIDDEN LIFE OF THE
WILD HERDS OF AFRICA

CAITLIN O'CONNELL

FREE PRESS

NEW YORK   LONDON   TORONTO   SYDNEY

*f*P

FREE PRESS
A Division of Simon & Schuster, Inc.
1230 Avenue of the Americas
New York, NY 10020

FREE PRESS and colophon are trademarks of Simon & Schuster, Inc.

For information regarding special discounts for bulk purchases,
please contact Simon & Schuster Special Sales at 1-800-456-6798
or business@simonandschuster.com

Book design by Ellen R. Sasahara

Manufactured in the United States of America

1   3   5   7   9   10   8   6   4   2

Library of Congress Cataloging-in-Publication Data

O'Connell, Caitlin.
The elephant's secret sense : the hidden life of the wild herds of
africa. / Caitlin O'Connell.
p. cm.
1. African elephant—Behavior—Namibia.   2. African
elephant—Namibia—Sense organs   3. Animal
communication—Namibia.
QL737.P98 O26 2007
599.67/415 22   2006052189

ISBN-13: 978-0-7432-8441-7
ISBN-10:   0-7432-8441-0

To all those who have lived and died in the name of conservation. And to William J. Hamilton III who passed away on April 24, 2006. His passion for conservation will always be a source of inspiration in my life.

CONTENTS

It's not what you look at that matters, it's what you see.

—HENRY DAVID THOREAU

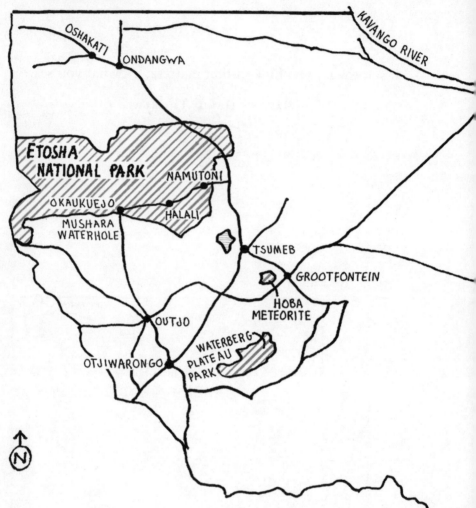

ANGOLA

KAVANGO RIVER

OSHAKATI

ONDANGWA

ETOSHA
NATIONAL PARK

NAMUTONI

OKAUKUEJO

HALALI

MUSHARA
WATERHOLE

TSUMEB

GROOTFONTEIN

HOBA
METEORITE

OUTJO

WATERBERG
PLATEAU
PARK

OTJIWARONGO

N

NORTHERN
NAMIBIA

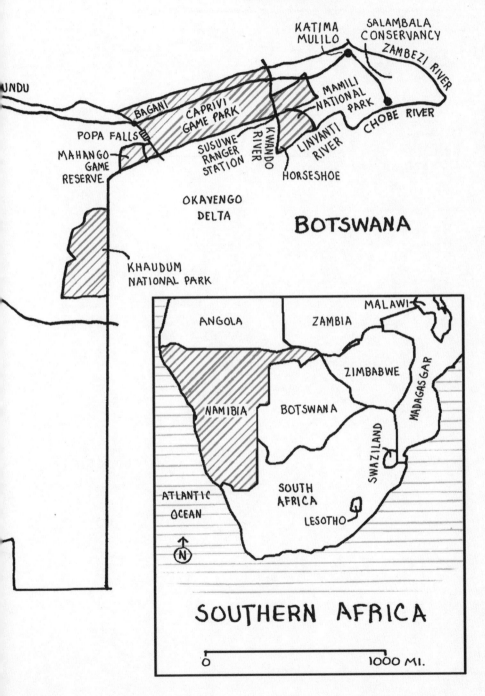

ZAMBIA

KATIMA
MULILO

SALAMBALA
CONSERVANCY

ZAMBEZI RIVER

UNDU

BAGANI

CAPRIVI
GAME PARK

MAMILI
NATIONAL
PARK

POPA FALLS

MAHANGO
GAME
RESERVE

SUSUWE
RANGER
STATION

KWANDO
RIVER

LINYANTI
RIVER

CHOBE RIVER

HORSESHOE

OKAVENGO
DELTA

BOTSWANA

KHAUDUM
NATIONAL PARK

ANGOLA

ZAMBIA

MALAWI

NAMIBIA

ZIMBABWE

MADAGASGAR

BOTSWANA

SWAZILAND

ATLANTIC
OCEAN

SOUTH
AFRICA

LESOTHO

N

SOUTHERN AFRICA

0   1000 MI.

1

LISTENING
THROUGH LIMBS

*Do not go where the path may lead, go instead
where there is no path and leave a trail.*

—RALPH WALDO EMERSON

I ALWAYS HATE TO LEAVE ETOSHA, but it was early August and
the winds were picking up. It was time to pack up the site and
head north. Sitting in the back of my truck, I looked through my
binoculars at the scrubby horizon dappled with giraffe necks,
trying to work up the energy to get my equipment organized
when I heard a thick, leathery, swishing sound right next to me.
I looked up to see 100 tons of pachyderm pass by, almost tiptoeing, heads bobbing in their Nordic Track–style gait. It was Broken Ear and her family, a group of twenty elephants, headed
purposefully toward the water.

I noted when Broken Ear arrived and watched as her family
assembled around the water. Being the matriarch, Broken Ear
occupied the outflow of the artesian well, which was controlled
in the dry season to sustain the large number of animals that
depended on it to get through this difficult period. The dominant elephant always got the best water.

Broken Ear took a few long draws, rolling up her trunk and

placing the water far back in her mouth, head tipped up. A few splashes escaped back into the pool. The howling gusts of wind had stopped, so I could hear the intermittent trickling of water escaping thirsty mouths. They were returning to the waterhole after their visit had been cut short the day before by the arrival of another family group led by a very intolerant matriarch that I called Collar.

Once the little ones had had their fill, they started to play. Young elephants are just as mischievous as lion cubs, always testing their boundaries with adults and jockeying for rank with siblings and other relatives. They stood at the pan's edge, knee-high in water, and tipped over onto their sides to make as much contact with the mud as possible. Then they began to wallow, swatting at each other with their wet-noodle trunks. Their mouths were turned up in what seemed to me like smiles as they wailed and cavorted in the water. The mothers merely looked on and continued drinking.

Reluctant to break camp, I savored the moment. I jealously guarded my visits to this splendid waterhole that the Owambo people called Mushara, after a tree common to this sandy forested habitat, or sandveld, a large terminalia with papery magenta pods (*Terminalia prunoides*). Mushara is located in the eastern corner of Etosha National Park, Namibia, one of a few waterholes restricted to the public, with the added benefit of having a protected cement lookout, which provided me with a great vantage for my elephant studies. I had a few other favorite waterholes where I would sleep, either in the back of my truck or in a hide. But I had developed a special bond with Mushara because of its remoteness and the number of elephants visiting on a daily basis. As much as I enjoyed the isolation, I was conscious of its potentially negative effect on my preception of reality. I was looking forward to being clean again, having not been able to shower for more than a week, but I was also dragging my feet about leaving the tranquility.

I was not looking forward to returning to my life in the north of the country, the Caprivi, where, despite its remote location, an exhausted community game guard might knock at my door at any minute, having bicycled 12 miles in the heat to assuage the anger of his fellow villagers to report another incident of a crop-raiding elephant. I would miss the elephants of Mushara, which were more isolated from the threats of war and civilization, and head to a place where they were different animals altogether. Many elephants in the Caprivi were still reeling from a war-torn past, exposed to land mines, automatic weapons, and poaching, partly to feed the hungry Angolan soldiers for the last twenty years. And at Mushara, I, too, could be a different person.

I first came to Etosha National Park in 1992, contracted by the Namibian Ministry of Environment and Tourism to study elephant movements, demography, ecology, behavior, and inter-actions with humans. My partner, Tim, and I were stationed in the Caprivi region of Namibia. This "day job" allowed us to spend our winters in Etosha, where Tim would analyze the movement data he collected from nine elephants fitted with satellite and radio collars while I would have the luxury of focusing solely on elephant communication and behavior. I hoped to use that research to help farmers in the Caprivi to pre-vent elephants from raiding their crops, which sometimes amounted to a whole year's worth of food consumed in one fell swoop.

It was during these off-seasons of my initial three-year con-tract that I was able to spend countless hours by myself watch-ing elephants at Mushara waterhole. I spent most of my time in the hide, a 10-foot-square cement bunker 33 feet from the water, with about 7 feet of its height buried in the ground and a pillbox slit facing the waterhole. It afforded a splendid view of elephants at close range but was very limited in terms of an overall picture of the negotiations made around the perimeter as to which herd was going to enter the clearing and when. But it allowed me to be closer to the action, and it was safe from lions or even a curious bull wanting to investigate the back of the truck. Once inside, however, I was committed for the night. It was me and my empty peanut butter jar, a makeshift but highly valued chamber pot in the bush.

I stayed in the dank bunker for as long as physically possible, usually about a week at a time. The isolation allowed me to reflect on the elephants' natural rhythms and to notice many behavioral patterns that had never been previously documented. These patterns formed the basis of my thinking for the next fifteen years of my career as a scientist. Eventually, once I learned to slow my own sense of time, adapting it to the deliberate, meditative pace of an elephant, I started to understand the patterns I had been observing.

Occasionally, Broken Ear would turn herself in a particular direction and freeze, sometimes with ears flat against her head and sometimes with ears held out, looking like a satellite dish, scanning the horizon. As she scanned, she seemed to cue the other adults to follow her lead. When she alerted the others to pay attention, they all oriented in the same direction, froze again, sometimes with ears flat, leaning forward, one front foot propped up on the toenails. Other times, they extended their ears, sometimes keeping toenail contact with the ground, sometimes lifting one front foot completely off the ground, swinging it front to back. It's not surprising that the matriarch would be wary at the waterhole and that she would be using her ears to try to detect any unusual sound, but what were these elephants doing with their feet?

I scanned the horizon and saw a large bull heading in on the southwest elephant trail, about a half mile away, ears fanning as he lumbered along. The vigilant cows shifted positions, several of them facing in the direction of the incoming bull. But their ears were not outstretched in a listening position. Could they have sensed his approach through the ground?

I became very familiar with this particular way that elephants seemed to be listening with their feet. My curiosity grew as I watched an elephant assume a position with its ears flat against its head, occasionally accompanied by leaning, with a foot lift or toenail contact. It was known that these animals used their ears like a parabola, scanning back and forth while remaining still to listen to low-frequency sounds from other distant elephants, but I had never read any studies describing an elephant's ears positioned flat against its head while freezing and listening.

Other scientists had thought that the leaning behavior was a resting position and foot swinging an act of indecisiveness. This is probably the case some of the time, but there were also some times where it appeared as if the elephants were actually using their feet to sense something in the ground. After seeing these behaviors, I would notice the arrival of a new herd or a lone bull or an approaching vehicle. When more than one elephant engaged in this freezing/leaning/shifting at the same time, it seemed as if the whole herd were using their *feet* to detect a signal.

I knew about the process of listening through limbs, which is found elsewhere in the natural world. Before signing on with the Namibian government, I had studied this phenomenon—known scientifically as seismic communication—in insects. In a sense, I was only going from the very small to the very large, having spent endless hours in a small soundproof chamber recording the seismic love songs of a unique group of Hawaiian planthoppers. It may be hard to understand my fascination, but I observed some extraordinary behaviors occurring on a little koa twig rigged to a gramophone stylus—behaviors that seemed very similar to those of these elephants.

When I would play the female planthopper's seismic call through the twig, the male would immediately press its weight down and would sometimes even lift up a leg or two, then inch forward, responding with its own seismic call, then freeze again, "listening" and perhaps attempting to locate the dummy female. I never could have imagined it sitting in that booth, but these little planthoppers and their extraordinary songs handed me a hypothesis for elephants that was so unexpected and controversial, I found myself occupied with the challenge of proving it over the course of the next decade.

Since many other species are known to transmit their vocalizations either through the ground or some surface such as a plant, it was not a completely ridiculous proposition that elephants might do the same. Many scholars have described how insects and other arthropods communicate by striking the ground or by coupling their vocalizations with a substrate. Other more complex animals such as fish, amphibians, lizards, snakes, and even crocodiles exploit this form of communication, too. However, the only mammals in which the mechanism has been well documented are small rodents such as the blind mole rat, the kangaroo rat, and the golden mole (though there is evidence that elephant seals may also share this skill).

Seismic communication may sound complicated, but it simply means a sound wave used for communication that travels within the surface of the ground as opposed to the air. Although it seems stiff, the earth is actually elastic. Tossing a stone across its surface, for example, would cause it to ripple, just like the surface of water. These surface ripples can be imagined on a larger scale such as in the case of earthquakes.

On a much smaller scale, surface waves from the slightest bouncing up and down of a hungry spider at the edge of her trap become a dinner bell. Similarly, a plant stem supports the same types of waves for a lonely planthopper looking for a mate. I discovered that the earth also acts as a sounding board for the elephant. But I knew it was going to be difficult to be certain and then convince others in my field that a large terrestrial mammal might be communicating seismically.

The concept isn't all that different from a person putting an ear to the railroad tracks to listen for a distant train, literally keeping an ear to the ground. It was illustrated in the movie *Dances with Wolves* when Kevin Costner listened to the ground to feel the tremors from a stampede of distant tatonka or American bison, a technique used by Native Americans to prepare for a hunt. Or, for *Lord of the Rings* fans, Aragorn putting his head to the rock to listen for the distant thumping feet of the fearsome Urukai as they bore Merry and Pippin away to Isengard.

A student of a Native American tracker once told me that in order to feel the pressure waves from the heartbeats of their enemies, the Apaches would hold the hairs on the backs of their fingers up to the windows of their enemies' houses. That's pretty subtle stuff. However, maybe it is not so subtle a sense for those who have a real use for it. There are institutes for the hearing impaired, for example, with special wooden dance floors designed to facilitate the hearing of music through feet.

Since all primates have the ability to detect vibrations through specially adapted pressure receptors in the hands, feet, and lips, among other places, it's not at all surprising that we have the ability to detect vibrations. If we really paid attention, or perhaps if those vibrations were a more important source of our long-distance communication, we would doubtlessly be able to detect vibrations much more acutely. In a quieter era, low-frequency drums or instruments like the didgeridoo might very well have been an important means for sending a message to a distant receiver through the ground. For those who need this communication channel, such as the hearing impaired, the area of the brain that processes touch, the somatosensory cortex, can even take over the function of the auditory cortex, providing that much more area dedicated to vibration detection.

For the elephants in the wild, I was beginning to think that perhaps seismic signals played an important role in their complex communication repertoire. Perhaps they, too, had honed their skills at detecting vibrations to better interpret their world.

THE LOOMING BULL came and went. He was old and in musth, with all the telltale signs of this hormonal state: a sappy secretion streamed from his temporal glands and he dribbled urine, which left a sheen of crusty green algae around his penis sheath. He did the rounds, searching for an estrus female. With each, he would test the cow's vulva with his trunk, inspect any urine patch in the sand, and roll his trunk tip to the back of his mouth to reach the duct leading to the vomeronasal organ, a special hormone-sensing structure that let him know if a female was in heat. With mouth wide open, he looked like a dirty old man, testing the tannins of the progesterone equivalent of red wine with his flehmen grimace. Nothing smelled interesting, so he sauntered off.

At this point, Broken Ear also decided it was time to leave. She stopped drinking and stepped away from the waterhole. She oriented toward her destination, emitted one long, low rumble, and waited there. Although two others rumbled in response, nobody appeared to make a move, and after standing still for about thirty seconds, she repeated her "let's go" rumble. After several of these rumbles, a few of the other cows finally decided to address this call to action and slowly lined up behind her. The group then meandered off to another foraging spot.

I watched Broken Ear and her family disappear into the tree line and then returned to my packing, occasionally scanning the horizon for possible new visitors. After coiling some wires and packing up my camera equipment, I watched a dust devil on the horizon and tried to decide whether it was going to blow through camp. While my eyes followed its course, I suddenly caught sight of a chalky gray mass lurking in the bush just beyond the waterhole clearing.

It was a new herd, consisting of a young matriarch and most likely her two younger sisters and their calves. They stood there for the better part of an hour, smelling the air with their trunks held high, mouths agape with anxiety, pitting thirst against caution. Could she have been waiting for Broken Ear to leave? I hadn't noticed that Broken Ear was not amenable to sharing her waterhole visits. Except for Collar, so named because she wore a satellite and radio collar as part of a movement study conducted by park researchers, elephants in this remote region of the park seemed to be accustomed to sharing their waterhole visits with at least one other herd. It was clear that no elephant wanted to contest Collar's challenges, not even by sneaking around to the overflow of the well that formed a flat pan of water.

Collar spent many of her visits simply holding off other elephants so that her family could have sole access to the water. Because of her tenacity, I often wondered how she managed to get enough to drink. And what had happened to her that she would view sharing as a threat when other herds appeared to have adapted to the routine? Another collared elephant in my more recent studies displayed similar behavior and became known as the "Collared Bitch" by my colleagues, who were most entertained by her defensiveness. Eventually, I decided that she needed a kinder name and began to call her Margaret Thatcher. It appeared that she was simply acting in the best interests of her family.

Seeing this new herd's reluctance to approach the water with a strange vehicle parked near the edge, I began to think that perhaps I was the cause for concern. At this point at the end of winter, most elephants in this lonely eastern region of the park would have seen the truck at some time during the season. Maybe these were stragglers from the west where all the natural pans had dried up.

They continued to wave their trunks high in the air as they stood in a tight cluster. I was upwind of them. An elephant, knowing the scent of humans, must have smelled me but probably did not see me. Otherwise, the matriarch would have shaken her head at me, objecting to my presence by flapping her large ears to make a loud cracking sound.

The herd finally got the courage to break out into the open in a hurried, stiff-bodied cluster, calves under bellies. A lingering dust plume hung in the still air behind them. Once they arrived at the waterhole, they spread out, rinsed the dust from their trunks with a scoop of water skimmed from the surface, and sprayed it out in front of them before starting to drink. It must have been a long time since they had had any water.

They were an unusually small group for this region of the park. The matriarch was relatively young, probably in her late twenties, judging by her overall appearance, height, skin condition, and the quality of her ears. Elephants are born with wrinkly skin and get even more wrinkled as they age. Their skin thins and gets more delicate, their hips and shoulders tend to show more, the temporal region just under their forehead sinks in, and some of the older cows even appear to grow longer rather than taller. Their ears get more frayed at the edges, some with large holes or slits. Some older matriarchs even have a "broken" or bent ear, where the cartilage at the top gets damaged over the course of up to sixty years of long-distance travels.

Although tusks are often used by researchers to identify individuals, they were a difficult characteristic in Etosha to count on, because a mineral deficiency in the soil caused them to break easily. None of the elephants there had very impressive tusks. Still, shape was helpful, particularly if one was missing or was unusually long or splayed.

As the afternoon began to slip away, I no longer had the energy for the tedious task of unearthing the buried microphone cables, packing up the rest of my gear, and getting ready to leave. Besides, I didn't like to be out walking around the waterhole alone in the late afternoon with so many lions in the area. Nor did I want to drive in the dark unless I absolutely had to. So it wasn't hard to convince myself to spend one more night at the site with the ever-enticing thought that another recording session would be the best yet.

It was an affliction of the occupation, wanting to record just one more breeding herd's vocal interaction as the elephants left the waterhole, or an aggressive encounter between a breeding herd and a rhino, or a lion roar, or an aggressive bout between two male rhinos. Perhaps this was the night that the hyenas would hunt down a kudu, loping around the waterhole with their demonic giggling after the terrified young buck. Or, if I was especially lucky, I might even record an elephant warning call that I could test out in the Caprivi farmers' fields to see if it would scare off crop-raiders. I had come to love recording the sounds of the African night, so it didn't take much convincing.

Often there was a sign just before something worth recording took place. If a jackal stumbled upon a lion, for example, it would rend the night with a sharp staccato barking that was very different from the normal undulating howls that jackals made in chorus. It was almost as if they were barking, "Lion! Lion! Lion! Ra, ra, ra. Lion! Lion! Lion!" Or, if you were vigilant and the lion was close by, you could hear a soft moan just before the lion initiated a full roaring session. There was enough of a gap between this noise and the real vocalization that I was usually able to turn the recorder on in time to capture a complete sequence of roars.

Having decided to stay, I turned my attention back to the waterhole. When the young matriarch began to leave, she stepped away from the water and rumbled softly. There was a return rumble and then another. From previous observations, I found that the time between the initiation of a "let's go" rumble and actual departure seemed to vary from moments to up to a half hour, depending on the herd's motivation level given a particular situation and what could only be attributed to the authority of the matriarch. This one was obviously more effective than Broken Ear at rallying the troops by getting a reply and a reaction so quickly. The rest of the herd reluctantly lined up behind her and slowly moved off in patchy single file, disappearing into the bush at the edge of the clearing.

The coordinated vocalizations that occurred during departure from a waterhole were very well synchronized, with little or no gap between them, as if the elephants were trying to create a single, continuous repeated call. I always took careful notes on these interactive calls during departure because I thought they might have some significance. Were the elephants specifically trying to create a longer signal, one that might be more easily detectable over long distances? And would it also facilitate better reception through the ground?

After this group left, I sat in the quiet late afternoon light, enjoying the stillness of the air and the waterhole with no animals present. I knew it wouldn't last though, so I didn't let myself linger too long before preparing for nightfall.

I packed up the back of my truck and prepared to descend into my nighttime recording station and sleeping quarters. While I was in the middle of settling in, two middle-aged bulls meandered in from opposite directions, both looking like they had come up from the southwest of the park. Elephants from that area were always easily identified because they were covered in a powdery white calcrete and looked like ghosts rising up from the horizon. Another sign that the season was changing—the eles were on the move.

Because bulls typically spend a great deal of time at the waterhole, I had to make sure that I had finished preparing everything I could before they arrived or else I would be stuck for hours without my recording equipment. As these two approached, they saw me walking to and from the vehicle. I purposefully stood outside the hide until they reached the edge of the water. I wanted them to feel comfortable with my presence so that I could continue to finish burying the wires behind the hide. Since I wasn't observing bull interactions at the time, I wasn't worried about my effect on their behavior.

I was slowly recognizing a few of the bulls and started to feel like they were my custodians, their presence providing me with protection against the unknown in the darkness of the night. These silent giants became a great solace to me, particularly the older ones and one very distinctive-looking one with huge wide-splayed tusks and a gentle manner. I wasn't alone here. Even when I lay awake at 2:00 A.M. listening to their slow and contemplative drinking, I felt like somehow we were keeping each other company, like they knew I was awake, listening. And although I had never read about bulls emitting vocalizations in the context of leaving, sure enough, often when a few old bulls left the water at night, they made a long, low rumble, one starting the chorus and another ending it, as if to say goodbye to their mysterious companion, though I knew full well they were talking to each other.

The bulls usually released some flatulence prior to the rumble, so the "let's go" included a high-pitched clapping sound before the very low rumble, probably because of the energy required to emit this powerful low-frequency call. These bulls would emit several rumbles while heading off into the bush to spend the rest of the wee hours dozing and foraging. Or sometimes a single bull would call repeatedly into the silence, and sure enough, a herd would arrive at the waterhole a short while later. It was as if he were signaling that the coast was clear, or perhaps it was a musth rumble but at the time, I didn't really know the true nature of this call, just that it was correlated with the arrival of a family.

The bulls tolerated my presence, as long as I kept to the far side of the hide, but I moved more carefully now, knowing that at any moment they could decide to be less gracious about sharing their space with me. The bull elephant is the most powerful animal in the bush, the true king of the jungle. I wanted to foster the benevolent dictator side of the beasts, as I knew that a tyrant could lie dormant just below the surface.

After I finished setting up and closed myself into the hide, the bulls decided to leave. It was as if they were entertained by my presence and now the show was over. Before going, however, they slowly approached, sniffing around the narrow pillbox slit, and along the sand I had shoveled over the wires. In general, they were only this curious during my first visits to the site or if they sensed that something was different, like when I had to reconfigure my setup or leave the site. Any other time, they usually ignored me.

One bull inspected the rock cairn that protected my microphone while the other stuck his trunk just inside the slit to smell me. He was very tentative, not wanting to risk damage to his trunk. Inches away from me, two hairy, dripping lips hovered inquisitively and sniffed heavily, scanning back and forth like an alien in a B horror movie. I held my flashlight tightly, prepared to tap the prehensile periscope away if necessary.

I often felt that if I did not allow them to explore the equipment, they would become resentful of this intrusion on their space and take it out on the microphone. I was only a guest at their bar and bathhouse, after all. Perhaps an anthropomorphic sentiment for a scientist, but it seemed like the appropriate thing to do, and so far it worked. Sure enough, after satisfying their curiosity, the bulls sauntered off.

Eventually, more elephants arrived, this time the very large extended family led by Left Tusker. They approached the waterhole boldly, parading into the arena bathed in the peachy pink splendor of the setting sun. There were thirty-five individuals in this herd, most likely two bond units made up of several families, including a tiny baby that did not yet have control of its trunk. They all hurriedly lined up to drink, except for the little one, which swung its head, trying to get momentum into its dangling, flaccid appendage. Futilely trying to imitate the others in the dexterous use of their trunks, the little one seemed confused that this thing would do nothing but swing uselessly in front of its face. Meanwhile, the mother kept a very close watch, tucking the baby under her belly every time it tried to stray in the direction of a young cousin.

An elephant's trunk is an extraordinary organ, containing more than one hundred thousand muscles, giving it remarkable agility. The San, or bushmen, with whom we worked in the Caprivi call the trunk a hand, because it is so precise in its movements and its ability to grasp. However, this little one's "hand" was effectively a lifeless, hanging arm, paralyzed from the elbow down; the ability to control its limb would develop in another month or so. In the meantime, the baby kept pushing it forward by moving its head forward, causing a pendulum effect down the length of the trunk. In frustration, it would then start swinging the trunk in circles, whacking itself with its limp propeller.

It suckled a while with trunk hanging to the side, just next to its mother's front knee. (A nursing elephant is an incongruous sight because the mother's teats are located in the chest area. It always reminds me of the elephant's close relative, the dugong, whose head and voluptuous chest poking up through the long sea grass led to the tales of mermaids.) The baby then dropped to its knees, trying to reach the water by bouncing its head up and down to get its yo-yo to touch the surface. The poor thing finally reached just a little too far and *kurplunk*—it fell like a rock into the trough, surfacing with a shockingly bloodcurdling scream.

The screaming engaged the emergency excavation crew, where aunts and sisters probed at the trough with stiff trunks, trying to scoop out the thrashing baby, while mother held the rubbernecking crowd of youngsters at bay. They eventually fished the baby out, and it stood soaking wet, sulking under its mother's belly.

Perhaps because of the baby, the herd was very anxious and wouldn't settle down after the unexpected dip. Herds are always especially wary when a new baby arrives, as it is particularly vulnerable to lion attacks. Once, in the Caprivi, we witnessed the aftermath of lions isolating a young bull out on the open floodplain in front of our hut. He had strayed just far enough from the herd to be in peril and paid the price that day with his life.

Most young bulls seemed to know the exact distance where they could feel safe from danger yet separated enough from the herd to seem liberated. This was especially evident when watching the arrival or departure of a herd at a waterhole. A preadolescent bull would often parade out in front and be the first to arrive at the water, sometimes demonstrating his dominance to the rest of the animal kingdom by chasing off all the other animals, even if only a flock of guinea fowl. He was also the last to leave, lagging just far enough behind to seem independent. But if he pushed it too far, he generally realized the error of his ways and panicked, running, trumpeting, and bellowing off after his family. Back in the security of the herd, he tucked in quietly, looking almost embarrassed, as if he had nothing to do with the previous scene. Herds seemed so familiar with this routine that they appeared unfazed by these riotous outbursts.

Once elephant bulls reach about age twelve, however, the matriarch and the other adult cows lose patience with these antics and unsolicited sexual advances. They kick the adolescent males out of the herd, where they are forced to develop new relationships with other bulls or roam on their own with other occasional companions. It was always clear who the dominant bulls were in these bachelor groups by the response of the other bulls at the waterhole. Upon their arrival, all the others would step backward some distance, clearing the way for the dominant male to have room at the head of the outflow, the freshest water. Once he got into position, the younger males would take their turn approaching him, placing their trunk in his mouth before repositioning themselves down the line. It was like watching mafiosi paying respects to the don. The older bulls, however, would just look on and drink, as if to pay their respects with a nod rather than the more supplicating action of the trunk greeting.

By the time the younger males are old enough to enter into the bull society, they are usually too big for lions to contend with unless they are sick. It's difficult to imagine that lions could penetrate a whole herd of breeding elephants to get to a baby, but at the end of my last field season at Mushara, we saw the possible remains of exactly that. Tim drove up from Okaukuejo late in the morning to help me pack up my site, and just as we were leaving, a young cow approached the water with something dangling between her legs. We watched as she waded into the water to wash herself, swinging her front leg back and forth between her hind legs to flush out the placenta. Her brow was wrinkled and her eyes turned down as if in great distress. If her baby was alive, there was no way she would have left it behind.

After washing herself, she slowly headed back into the bush. We tried to follow her, far enough behind so as not to upset her, but then the bush got too thick and we eventually gave up. We wanted to solve the mystery. Was she heading back to her dead baby? Did she fall behind the herd while giving birth and the baby was taken by lions? We would never know, although it wasn't normal for a herd to leave a birthing mother alone. Usually, the whole family made a circle around the mother and waited for the baby to be born; and a baby elephant is born to run, so there isn't much waiting around afterward.

The elephant herd at the waterhole was now radiant and almost cartoon pink in the light of the setting sun. But they were agitated. In fact, I had never seen a herd so nervous. I had settled into the bunker for the night and couldn't see behind me, so I couldn't tell what was upsetting them, but there was clearly something.

My suspicions were confirmed. A small herd of gemsbok (oryx) reached the edge of the clearing and remained there, stamping their feet and snorting in objection to something in the bush behind me. I wanted to lift the heavy iron hatch of the bunker to check, but it would have made too much noise and I did not want to scare the elephants. I set up my video camera on a tripod and let it run.

Finally, seemingly out of nowhere, the matriarch bellowed. It was a deep, low, urgent rumble, the most impressive call I had recorded to date. After three such calls, she turned and fled, leaving the rest of the herd screaming and trumpeting behind her, the little ones trying to keep up, tails straight out, trunks flying. Not much of a coordinated departure this time.

Several minutes later, they reached the edge of the clearing, and then the waterhole went silent again. The twilight filtered through the thick cloud of dust hovering about a meter above the ground. I sat there in awe, waiting for an explanation. The pan was now a brilliant red, too dark to film, so I turned off the video but left the tape recorder running.

Meanwhile, a covey of double-banded sand grouse trickled in and filled the edge of the flat pan, the spillover from the trough, stashing precious droplets in their feathers to take back to their nests. I sat and took in the still but melodious scene as more and more grouse spilled in. There was something else out there. The feeling was palpable.

If you are very focused, you can sometimes smell lion. It's a kind of thick, leathery smell, similar to a buffalo but more musty. I don't know whether I wanted to smell them or whether I really did, but I knew there must have been some lions sitting at the edge of the clearing behind me. I waited for something to happen as the grouse had their fill and left behind an unsettling silence.

The answer to the mystery was soon revealed with an over-powering roar that pounded on my chest and quickened my heart with each bellow. Then came a reply from the adjacent corner, and then a volley of heaving roars as the two cats approached the water. I got out our government-issue Zeiss low-light binoculars and scanned the area, waiting, frozen in my stance, marveling at how assaulted I felt by the tremendous pressure waves of their roars. The chorus crescendoed. I knew I would get a fantastic recording.

Suddenly there was silence again. Then I heard the telltale sign of a large predator at the waterhole: the slow beating of a weighty tongue as it laps at the surface of the water. I heard it off in the distance, and then there was another, even closer lapping. Finally, I could see them: two young males sporting barely respectable manes, but they were very large, and they meant business. I settled in, excited at the prospect of the amazing record-ing I would capture from this potentially active night. But eventually, overcome with the exhaustion of waiting, I passed out.

I awoke early the next morning to the sounds of vultures bickering at the waterhole. Groggy and stiff, I shielded my eyes from the blinding light of the outside world and peered through the slit. It was a gruesome sight. Vultures lined the water's edge, sticky with blood; more were perched in a tree at the edge of the clearing, and two marabou storks adorned an adjacent tree. There was not another animal in sight.

I climbed up the ladder of the bunker and carefully slid open the heavy manhole cover so as not to upset the resident geckos. Standing on top, scanning the clearing with my binoculars, I could see more vultures directly behind me. No sign of the lions but the remains of their dinner, a young male gemsbok, was plainly visible. At this point, the only thing left to identify him were his horns and the black and white socks still left on his ankles. The rest was bone.

Knowing that the lions must still be in the vicinity—they wouldn't stray far from a kill or the water—I did not want to pull in the wiring from the waterhole without the truck, so I started to pack up the other equipment. I carried a big stick as I walked over to the truck to bring it closer to the hide. Since I had not expected to stay another night, I hadn't charged all of my recorder batteries the day before and had to tap into the truck battery to record the lion roars. One turn of the ignition made it clear that the battery was dead. I took the solar panel out of the back of the truck, placed it at an angle on the hood using a rock, and attached the battery. Judging from the sound of the alternator, it would take a few hours to charge.

While waiting, I pulled up the wiring between the hide and the truck, keeping careful watch for the lions. They were probably resting their gorged stomachs in the shade at the edge of the clearing somewhere, so I took comfort in knowing that they wouldn't want to bother with me in that condition. I still kept an eye out as I packed up the rest of the equipment.

When the battery was fully charged, I put the last few bags into the back of the truck and the recording equipment in the front. I jotted down some last notes on recording improvements I could make for the following season. I wanted to see if the low-frequency, high-amplitude acoustic rumbles that elephants made were capable of moving through the ground as well, which might explain the curious behaviors I had been documenting.

I had a lot to learn about recording sound in the ground. I knew that seismologists used a geophone to measure earthquakes, which worked pretty much like a condenser microphone with a magnet and coil. The movement of the earth moved the magnet and coil, which generated a voltage that scaled with the strength of the signal. But how many geophones would I need and how expensive were they? And how would I amplify the signal? With very little access to lines of communication with the United States, it was going to be all the more difficult to develop a discourse with potential colleagues in a field about which I knew very little. I could see Mount Everest standing in front of me, but I was still excited about the possibilities.

I said my good-byes to Mushara and headed north, back to the Caprivi. It was a long eight-hour stretch with not much distraction to stay awake, other than the sudden shock of a herd of cattle sitting in the middle of the road while barreling toward them at 75 mph or a lone kudu leaping out in front of the truck. Needless to say, it was much safer to travel the country by daylight than by night, but that wasn't always possible.

And yet the eight hours was barely enough to transition from my solitude in the semidesert expanses of Etosha to the green and lush but tumultuous Caprivi. Our supervisor, Jo Tagg, liked to call the Caprivi "God's Country." And yet he'd say that survival there required "honing one's ironic distance," essentially meaning that you needed to get good at not letting anything affect you at any level.

The Caprivi had been a war zone not just between humans but between humans and nature, and farmers were pitted against the elephant in a battle for survival. The tragedy was that its striking beauty went unrecognized by those who could make a difference and was voraciously consumed by the rest.

2

THE CAPRIVI

*One of the elephant's favorite foods, the Acacia
erioloba, is thought to attract lightning, and a potion
containing charred portions of the lightning-struck
tree, mixed with goat's fat, is used as a protective
charm during public debates.*

—VERONICA ROODT, *The Shell Field Guide to the Common
Trees of the Okavango Delta and Moremi Game Reserve*

THE CAPRIVI SITS IN NORTHEASTERN NAMIBIA, a narrow
strip of land connecting with Botswana to the south, Zambia
and Angola to the north. The vestiges of German colonialism,
the area started as a makeshift trucking route across southern
Africa, striking, raw, and untamed. Yet it is a vulnerable region,
too, exposed like an unclaimed gem in the desert. After World
War I, the Germans surrendered their colony of Southwest
Africa to South Africa as a protectorate, and the narrow strip
remained an unfinished dream.

At the southern tip of the Caprivi lies the Linyanti swamp
above the confluence of the Kavango and the Kwando Rivers,
just above the Okavango delta. It is a place where elephants still
wander the land freely, returning to where their ancestors have
migrated since time immemorial. They are free to revisit the
migration routes, the fruit trees, their favorite acacias, and the

bones of their ancient battlegrounds. Sisters, brothers, aunts, and cousins, they were caught in the middle of men who fought bloody wars for the sake of liberation, remembered now by scattered wooden crosses in the forest, suggesting for the moment at least that it might be safe to wander here. Even without physical boundaries or landmarks, they somehow knew that farther north the slaughter among men continued, and with it, the danger of their being butchered as well by desperately hungry troops.

At the end of the dry season, elephants fill the floodplain, a thousand strong in one continuous gray streak, stretching across the golden turpentine grass, rumbling, trumpeting, and screaming jubilantly under a threatening sky. The tusks of young bulls clack like billiard balls as they joust their way across their domain in a game of dominance for future wars of their own. As the clouds build with the promise of rain, the elephants scatter to the four corners, each herd in search of its favorite wet season pans.

We lived in an area known as the Triangle, a remnant of the South African Defense Force (SADF) that was not designated as part of the rest of the West Caprivi Game Park. The whole area was littered with ramshackle military installations, remnants of failed negotiations between the Department of Nature Conservation (now called the Ministry of Environment and Tourism, or MET) and the SADF after Namibian independence in 1990.

The Triangle, a lush riverine environment, contained the highest concentrations of game in the West Caprivi, but it had military immunity, making it invulnerable to the laws of the park surrounding it. The designation was never rectified after independence, leaving the area unprotected and the source of many failed economic schemes. Once, nine Bulgarian tractors were airlifted in and off-loaded at the Katima airstrip, a political gift for the president after independence. They were to be used to plow the Triangle for cornfields. Fortunately, things worked out in favor of the thirty-five hundred elephants and one thousand buffalo. African momentum proved mightier than the plow, and the tractors never broke ground. They never left the airstrip.

In the Caprivi, violent death is as much a part of the landscape as the capricious nature of rain. Nobody knows when it will come or how much to expect, but in the end it always comes. Death can snatch people away without warning—for example, a leopard stealing into a hut leaving a faceless victim, a croc seizing a laundress off the riverbank, or an elephant using its powerful knuckle to smash the ribs of a hapless person lost in the forest. A lone male buffalo may crush his victim with the iron-hard boss of its horns; when wounded, it can unnervingly circle back to kill the intruding hunter with a surprise attack. And a neighbor may disappear simply for being from the wrong tribe, or from the cold sweat of the ever-present malarial fever, or even from an unexpected twist in the night, silencing the cries of an infant.

Wildlife, humans, and disease are not the only random slayers. The Golden Highway in the wet season delivers an untold number of victims to their deaths. The chalk-white greasy puddles of this main thoroughfare across the West Caprivi turn the pavement into a sheet of ice.

There were secret gems here, too. Hidden beneath the desperation, they crept up on me with conflicting emotions of love, fear, need, and rejection. The undulating apricot-colored dunes of the Namib pulled at me, filled my dreams with sliding sands and jagged, stark edges, haunting me with contrasts. Light and dark, wet and dry, beautiful and wretched, peaceful and warlike, Africa always presented two conflicting aspects to her cruel yet beautiful face. She was like the stories of the leopard that the laborers in Kruger used to recount: If you challenged a leopard, they warned, looked at it in the face, it would disembowel you. If you let it alone, it would avoid you, the only evidence of its presence a territorial, rasping call in the night.

Suppressing my fears and seduced by this overwhelming

beauty, I resolved to take the position that the government had offered Tim and me to study elephants in the Caprivi. When I was at my best, I could see it not so much as an impossible mission but as a chance to make a small dent, to sow a glimmer of hope for the poor farmers who had reached the end of their rope with the crop-raiding elephants. And when times were good, Tim and I reveled in our bush lives, our little thatch hut next to the Kwando, a tributary to the great Okavango delta in Botswana, teeming with elephants, buffalo, hippos, antelopes, and spectacular birds and insects.

But despite the enthusiasm of our super adviser at the time, Malan Lindeque, there were some strong prejudices about hiring foreigners, due to fears about being misunderstood and thus misjudged internationally. Tim grew up in South Africa speaking Afrikaans, and, ironically, that was far less offensive than my being an American. There were strong anti-American sentiments, particularly since the U.S. Agency for International Development (USAID) had just arrived with a package to sponsor several rural development programs but did not make a very good first impression. At the end of the initial meeting with local development workers, a USAID woman got up and said, "Well, by the end of our projects here in Namibia, we hope to see a lot more native Namibians involved." A Namibian woman, Margaret Jacobsohn, with a long history of community activism in Namibia responded, "Shame that I don't notice any Native Americans on your team." It was not a smooth introduction.

Tim and I hadn't had any prior experience working with elephants other than volunteer time spent counting them in Etosha National Park. I had a master's degree in entomology and Tim had a bachelor's degree in biology. We had come to Africa for a break between degrees, with the original goal of taking a year to drive from South Africa to Kenya. So a three-year contract offered by the Namibian Ministry of Environment and Tourism to study elephants in the Caprivi sounded like a pipe dream, but such was the haphazard nature of Africa. The closest bird got the worm.

After two months of volunteer work in Etosha, Malan, the head of research at the time, took a liking to us. One thing led to another, and because the first researcher ended up working with lions instead of elephants (after waiting two years for the money to come through from the European Union), the money sat there with no one to do the research. Because there were two of us, Tim was to study movements and demography, and I would focus on elephant ecology, behavior, and interactions with farmers. They got a two-for-one deal in hiring us.

When one year was left on our contract, we contemplated extending our stay, but felt that we'd have more to offer later with Ph.D.'s in hand. Tim's job was evolving more and more into a technical position and, along with the elephant tracking, he had set up the aerial censuses for the region. He had just finished compiling census data from the past ten years to add to elephant and other important large mammal counts of the past two years. He had a lot to offer, but we still felt that we needed more time to focus on the scientific method before fully immersing ourselves in a managerial role.

Outside of the magical time I got to spend with the elephants in Etosha, I was becoming increasingly involved with the women of the Caprivi, helping them with resource management and community development. The conflicts between farmers and elephants were just the beginning of the problems I would face, and ultimately a reevaluation of the spirit of the community and its economy proved to be the most effective sector for my energies, as community attitudes and perceptions of wildlife in general played a critical role in decisions made about elephant habitat and conservation.

At that moment, the government was in the midst of a profound transformation. In the past, wildlife was government property, where farmers experienced only the problems resulting from living with these animals and none of the benefits. Land reform legislation was in the works to attach a monetary value to elephants so that farmers would no longer look at them as competitors. Local communities could now "own" the wildlife through conservancies, so they had an interest in protecting it. This encouraged local communities to form conservancies, allowing them to benefit directly from revenue generated through ecotourism.

But those were long-term solutions, and I needed to do battle in the here and now. Farmers and elephants in the Caprivi were both victims of circumstance, their fates inextricably linked through the competition for land, food, security, and access to water. In desperate times, violence was often seen as the only course.

The Caprivi human community was insular, and we remained on the outside for some time. It was particularly harsh in the beginning, punctuated by our first meeting with Jo Tagg, the administrative overseer of our project and acting chief nature conservation officer for the region. A young woman showed us into his office, her English South African accent and freckled face warming the soulless building. She introduced herself and said that Jo would be back shortly. Then we waited, nervously assessing the disheveled office of our prospective supervisor.

The faded walls and sagging map of the Caprivi bore down on me. It was as if the entropy of Africa sat in the bloated, once waterlogged ceiling boards, now shrunken and turned up at the edges. An army of large wasps patrolled lazily between the exposed metal framework that served as walls. Things seemed different now that the job was almost a reality. The romantic images of a little thatch hut along the Kwando River were soon replaced by the harsh realities of a newly independent African country. The empty supply warehouses, empty army airstrips, and skeletons of large army trucks cluttering the park headquarters served as reminders of a complicated war for Namibian independence.

We heard the roving sound of a land cruiser, a door slam, and then faint muttering from a man approaching the building. The girl tittered in the next office in response.

She whispered, "Some people here to see you."

Silence. We looked up at the open door to see a tall, balding man lurking in the doorway wearing an untucked Ministry of Nature Conservation shirt. He was kneading a floppy hat.

Jo paused. "Is it?" He searched the girl's face for the nature of the visit while trying not to look at us.

"Young white couple. The girl is American," she confided.

Jo rolled his eyes, cursing in Afrikaans. "Oh Je'sus."

He entered the office; we struggled out of our chairs. "How's it," Jo thrust his wrist down at Tim, his hand limp. An almost leering expression slipped off his face when he glanced at me. He was caught off guard. He knew that I had seen it, but cleared his throat and skid past us to sit behind his enormous desk. A safe distance. "Jo Tagg. What can I do for you?"

We introduced ourselves as Jo held the desk in both hands, arms stretched. He rolled his eyes back and locked onto Tim so as not to slip up again.

Tim explained our purpose, our past, our recent experience volunteering in Kruger National Park, South Africa, and Etosha, and that Malan had wanted us to take the elephant research position in the Caprivi and suggested that we stop by to see him. He glared at my kneecaps as Tim's monologue turned from conversation to confession. I instantly regretted having worn shorts. Tim continued, hoping that Jo could shed more light on the project.

Jo played with his pencil. "Where are you from? I recognize your accent."

"Natal."

"Good God, mate, a Natal boetie! I'm from Durban, myself. Welcome to God's country!"

"And Caitlin's from the States," Tim offered.

Back to staring at my knees, "Yes, America. Home of the brave."

I tried to break his stare by crossing my legs. I perched there like a springbok surrounded by a pack of giggling hyenas. He couldn't hide it. I prompted him by asking again how best we might succeed in such a position.

"Hopeless. Absolutely hopeless. You can't save the elephants here." He winced as he crossed his legs. I could tell that he thought the bush would eat me alive. He grimaced and pulled at his shorts. "European Union funding. Ha! Their biggest concern is what diseases our roaming elephants can spread to their cash cow in Botswana. Clever disguise for a plan to fence the whole southern border, I suspect."

Tim tried a different perspective, suggesting how we might get a better understanding of their movements and demography, and how that information might help protect migration routes in the future.

"All I can add in my solitude is, may heaven's rich blessing come down on everyone, American, English, or Turk, who will help to heal this open sore of the world." He paused expectantly. "David Livingstone's last words. They're inscribed on his tomb in Westminster Abbey. I find them quite soothing."

He flushed and continued, "Though leaves are many, the root is one; through all the lying days of my youth, I swayed my leaves and flowers in the sun; now I may wither into the truth. 'The Coming of Wisdom with Time.'" He squinted at me. "William Butler Yeats. Do you like him?"

"Very much."

"There are a few good things that come from America. Know any Robert Frost?"

"Have you been to America?"

"No, but I plan to. In fact, I am going over to save America! A lot of you Americans come to Africa thinking that you are going to save Africa. Save Africa! Ha! Well, I decided that it would be far more productive for me to go to America to save America. That might get us closer to saving Africa, don't you think? Quite a pragmatic approach, I thought."

"Pretty ambitious goal." I tried to humor him.

"And noble!" Jo chortled.

I felt as if I had broken through to him for a split second.

Jo gave me a cursory glance up and down to see if he had missed something and decided to ease up. He leaned forward, stretching his hands over the domain of his desk. "Elephants are the most important resource this region has to offer. It is bloody important to find out more about them, where they move, how many are poached, age distribution, mortalities. And the elephant problems are ruining our conservation efforts. No benefits, just costs. These people must start seeing benefits from wildlife, not just the damage to their crops."

After a moment of silence, he suddenly got up and tugged down his shirt jacket. "Well, it was a pleasure chatting with you. I'm afraid I have an appointment with Chief Mamili just now. If he had his way, there would be no bloody pachyderms left for you to research, anyway." He laughed, thrusting his wrist into Tim's lap. "That's one way to simplify your study. Let them eat meat!"

Tim struggled to get up and confront the limp wrist head on.

"Oh, and one last thing." Jo giggled, "I hope neither of you has a sensitive stomach. The Caprivi causes ulcers." Jo waved us out of his office.

We got back into our piping hot '73 VW Bug, outfitted with our poorly welded homemade roof rack constructed in Tim's mother's garage, which held gas cans, an extra water jug, and spare tire. We turned left toward the Chobe, crossed the Botswana border, and as we put distance between us and the strange encounter with Jo, we looked forward to seeing Victoria Falls. Since we wouldn't start work for another three months, we aimed to get to Nairobi to finish our original travel plans while we waited.

Having taken in the falls, "The Smoke That Thunders," from both the Zimbabwean and the Zambian sides, and having survived the potholes of Zambia, we spent another few weeks on the tranquil shores of Lake Malawi, watching the fishermen haul in their nets at sunset. We were turned around at the border of Tanzania for not having a *triptique,* a considerable deposit on our vehicle to deter us from selling it in East Africa. Apparently, the border post was operating under a different set of rules than the embassy. By that time, we were tired of being on the road and eager to get the project started, so when we got the telex that we were officially offered the position, we drove straight back to Windhoek to sign the contract.

I stocked up on electronic supplies such as a GPS, laptop computer, portable printer, VHS radio tracking system, and sound recording and video supplies, all ordered in the U.S. while Tim handled buying an old Toyota Hilux, outfitting it with a long-range gas tank, bull bars, and a cage on the truck bed.

By the time we got back to Katima, Jo had had a complete change of heart about us. He greeted us with enthusiasm that was difficult to find sincere. We welcomed it nonetheless.

Jo had been up all night conducting an ivory bust and apologized for his exhaustion. He said he'd take us on a short tour of the town before he had to get back to contend with the "evidence," a pile of muddy elephant tusks and a clutter of animal parts and skins in the corner of the office.

Jo introduced us to the capital of the Caprivi, Katima Mulilo, with an amicable running monologue, first bumping into Kai, the conservation officer for the region, who stopped in to say hello. He was wearing bright green ministry overalls, which, by the look of them, had been dipped in battery acid. He had a rugged, sluggish look, was friendly enough but curt, and though his fingers were meticulously clean, his cuticles were black.

"Kai, how you?" Jo signaled for Kai to have a seat and gave us one-liner introductions before inquiring about his latest exploits.

"I heard you nailed the Lizauli croc today! Nice. Five meters and a piece of her dress!" Jo winced.

"Je'sus." Kai cursed, pronouncing "Yesus." "Jo, you should

have seen it. The mother of all crocs. I didn't think we'd find anything inside." He shrugged with an accent mixed with German and Afrikaans.

"Nice work, Kai. Were the people happy?"

"Jera those Lizauli mensa, Je'sus! They burned the skull. The brains are poison!" Kai shook his head. "These people. Witchcraft will be their undoing! Do you know there is a Zambian witch doctor that is so rich he flies around in his own airplane with his own pilot! He is charging these people a fortune. I heard he charged a guy twenty head of cattle for special medicine to make the guy's wife pregnant. And then he required that he sleep with the wife for a whole week as part of the treatment. Do they really believe this stuff?"

"Never underestimate the capacity to believe."

"He has probably sired two hundred offspring by now."

"Bad subject at the moment," Jo grimaced in a whisper, spreading his fingers wide and holding his hand out. "He's been implicated in our latest bust." His hushed voice cracked with restrained excitement. "You know the five remaining that we were looking for?" He beamed, raising his woolly brow. Kai widened his eyes. "Well, there are supposedly fifteen more. It's all coming from his place through the Singalamwe route." He tilted a hand down again to close the subject. "We'll discuss it tomorrow night at my place. Big plans for next week," he disclosed, squinting his conspiring eyes as he paused and changed his tone. "And the flat dog skin?"

"I salted it. It's at the Nymbwe office. You know, it's really not well that Treasury claims the skins. At least the family could have gotten something out of it, funeral costs if nothing else."

"Ha! Good luck!" Jo looked at his watch. "Oh, hell. I've got to quickly run out and show these two the hardware and the wholesaler. Any tips on bricks?"

Tim was hoping to have our field station built as soon as possible in order to beat the rainy season. Kai generously offered his most valuable resource, the brick maker. "We have to make the bricks?" Tim smiled nervously as I tried to remain expressionless.

Excited by his mission, Jo wanted to get the tour going and insisted that he drive. He got in and cleared the mess off the front seat to make room. Tiny colored globs of candy oozed out of cellophane packets on the dashboard.

As we backed out, another young man pulled up next to us in a turbo-charged white Land Rover, a large British flag painted onto the passenger door with the slogan "Donated by the British Government." He got out and greeted Jo. His khaki ensemble was also spattered with battery acid holes. Jo called jovially, "Matthew! How you? Are you well?"

"How's it, Jo?" Matthew replied, shaking Jo's hand and clutching an empty pipe. "How did the khuta meeting go?"

Jo removed his sunglasses to display his weepy, theatrical eyes and rubbed his hand down his brow. "Hopeless! Bloody hopeless!" He squinted up at Matthew. "Listen, Matthew, I've got some people I'd like you to meet." Matthew ducked down to greet us and we all exchanged nods.

"These are our new elephant researchers," Jo explained with suppressed sarcasm. Matthew seemed surprised.

"Well, they'll be around for a few years, so I'm sure you will have plenty of time to talk. Matthew here runs the community game guard program for the East Caprivi for an outfit called IRDNC. He can tell you what it stands for. I can never remember."

Matthew nodded. "Integrated Rural Development and Nature Conservation." Jo put the truck in gear.

"Right, now if you'll excuse us, I'd like to show them a few places around town before this afternoon's trial. See you later."

"Nice meeting you," Matthew called as we sped away in the pickup, bumping down the potholed track.

The Caprivi seemed to be full of young men in positions of power disproportionate to their age. For government positions, a Caprivi post was a way of doing time, a stepping stone to a higher position elsewhere. For foreign aid positions, it was a convenient way to skip the ranks and run the show. Nobody wanted to be posted to the Caprivi. An unfortunate thing in an area where negotiations were fragile, steeped in a complex cultural tradition where seniority was proportional to age and wisdom. I hoped that he was nice beneath his pompous veneer.

Jo offered me a melted candy. He rattled off a joke about Matthew's newly donated vehicle and how the Land Rover was a thinly veiled disguise of British revenge for having lost the colonies. He popped a sticky green gob into his mouth, which stuck to the roof of his mouth as he directed our bouncy tour through the dusty town square.

"Ah, we must hit the hardware first. How are you at baking black forest gâteau?" Jo nudged me. "We must get you a little oven to make nice black forest gâteau. One must be civilized in the bush."

We pulled up to the hardware store and marched in. He led us into the immense but almost empty, dusty showroom. Tim focused on the plumbing aisle, fitting parts together and mentally planning the logistics of our new home in the bush.

I drifted off to another aisle in search of something to steel me against a wave of panic. I looked at Cadac stove fixtures, then the cast iron pots. There was nothing to comfort me. All eyes in the store were fixed on me, a curious anomaly. Should we have just gone home? Traveling with few resources was an adventure, but trying to work with few resources was starting to seem more than intimidating.

I walked toward the back of the showroom, where a locked fence gate separated the stockroom in the back. Jo and Tim had gone in. I found someone to unlock the gate, and, following the sounds of Jo's giddy narration, I caught up with them.

"Booming new business," Jo pointed to three men assembling coffins in a corner. "Some say that white man created AIDS to get rid of black people." Jo ran his index finger down the length of one of the coffins. "Some say it's witchcraft. Others don't believe it's real. Dr. Gous reckons there're at least three a day, and the toll is rising." He looked down at his watch. "Come, I'll take you to the wholesaler. We'll have to see the clinic another day."

As we got back into the truck, Jo shooed away a dog with sickly yellow eyes. "Street rats. They run in packs all night. The unfortunate thing about putting your head on your pillow. It's like an alarm that sets them off. Horrible creatures." He left the dusty square and headed down a tarry section of road. He just missed a dog as we turned a corner. "There have been so many dogs at times that finally the mayor ordered all dogs to be taken off the street. There was a twenty-four-hour grace period before the leftovers were shot. It worked well for a while, but I think it's time for another roundup."

Jo parked under a tree next to a few more mangy dogs panting heavily in the shade. "Needless to say, Kai set the record for shooting the most!" Jo laughed as we entered yet another dusty skeletal building. "Oh, yes, uniforms. We need to get you into uniforms as soon as possible."

A young man in a bright Salvation Army–style shirt approached Jo slyly. He wore oversized dark sunglasses and a bead of perspiration above his upper lip. Jo stiffened and whispered insistently, dismissing the young man. The boy ran toward the exit before Jo could swat at him.

"Damn it!" Jo cursed as he continued down the aisle, trying to regain composure until we reached the canned corned beef. "Yes. Uniforms. Very important."

Just before our arrival, a postal worker had been shot in a nearby village because he was mistaken for a Ministry of Nature Conservation employee. "Wouldn't farmers have a negative reaction toward me if I approached them about their elephant problems wearing a uniform?" I picked up a can of corned beef and dusted off the lid.

"You must be in uniform!" Jo stormed. "The ministry must get credit for this." He pointed at me. "This is a *ministry* project!" He paused, trying to work my eyes off the floor. "The communities need to see something positive coming out of this ministry," he softened.

After I mumbled something sufficiently acquiescent, Jo continued the tour. "Coffee with chicory," he grimaced. "The scourge of Africa!"

When we got back to the ministry office, we thanked Jo for the tour, and he waved us off with some parting words of wisdom.

"Go well," he said while shrugging his brow, sensing the enthusiasm of newcomers. "And don't get your hopes up. God's country is full of broken dreams!"

3

WET SEASON
ELEPHANT WARS

*He setteth up his tail like a cedar, the sinews of his
testicles are wrapped together. His bones are like pipes
of brass, his gristle like plates of iron. He sleepeth
under the shadow in the cover of the reed. Behold!
He will drink up a river and not wonder that the
Jordan may run into his mouth.*

—JOB 40:10–18

I SAT ON OUR PORCH AS THE CLOUDS came to a boil over the
floodplain. A strong wind preceded the rain, rolling the grass
down in waves, the tops of the reeds a deep gray, tickling the
dark purple sky. The sleek gray of a leadwood carcass glistened
in a distant tree island while all around it the light intensified the
yellows and reds of the grasses. A chalky blue elephant herd
hurried with heads bobbing in a train to reach the tree line
before the storm broke.

A furious flash of energy escaped from the sluggish sky.
Thunder broke deep and low with such palpable weight it
seemed to crack open the earth to lay bare the fiery rock below.
The ground shook as the battle over the floodplain continued

until an all-encompassing torrent of rain smothered both thunder and lightning. The remnants of the battle steamed and hissed in the downpour.

In our first storm in the Caprivi, Tim had immersed himself completely in the earthly forces. Tearing off all of his clothing, he ran out into the downpour, yelling at the roaring sky, "Let slip the dogs of war!" Drenched and thoroughly exhilarated, his body silhouetted against the electric sky, he screamed at the top of his lungs. Yet his voice was so completely absorbed by the raging storm that I could barely hear him as I stood watching his primal display from the porch.

We were staying in the peeling, dilapidated Susuwe camouflage barracks at the time, the house slated for Jo. At first, we were hoping to build before the rain, but as soon as the rain started, momentum stopped. Time runs on a much slower clock in Africa, but living out of boxes was starting to get old. And I didn't like being in that house by myself. When the second storm hit, Tim was stuck in Katima for the night, and I was forced to face the earthly forces on my own. In retrospect, it seems that I had no problem camping in the bush on my own, but there was something creepy about that military barracks. Too much history, too many failed politics and past haunts, such as the shotgun hole in the bathroom wall, allegedly from one of the rangers shooting at his wife. And the nightmares from mefloquine, an antimalarial drug, didn't help.

Sitting under the tin roof, eardrums saturated with the roar, I stared unfocused into the wall of water that quickly accumulated and poured off the broken gutter, beating down on the orange sand. I sat in the downpour until it got dark. When the storm slowly let up, the noise of the torrent was replaced by dripping vegetation and the chaos of frog calls.

WE ADAPTED QUICKLY to our life in the Caprivi. The rain came and went, bringing plenty or drought, and the Zambezi swelled and shrunk accordingly. It was the only place I knew of where people measured their lives on meter sticks at the river's edge, the rising river an indicator of local rainfall. The breadth of the great Zambezi during the wet season represented that year's economic fate—riches or disaster. Cataclysmic and deadly encounters were ushered in by thunder and lightning, and the pestilence of the wet season meant that the place was abuzz with life, renewal, and parasites.

Sex was in the air, in the water, and all over the ground: tok-tok beetles banged their shiny, black abdomens against the earth to attract a mate; red velvet mites scurried about, the enormous females chased by scores of tiny males. The tall red and orange turpentine grasses were in bloom, too, a fragrant bouquet filled with all variety of grasshoppers and mantises. A great number of these ended up on our windshield during our patrols; the grass seed choked the radiator.

Emperor dragonflies cruised the riverbanks, dining on the latest mayfly emergence. Baby tortoises crawled across the road. *Charaxes* butterflies hatched out everywhere, lining the edges of the wet-season pans with orange and black brushstrokes. Bee flies came to show off their fuzzy, swollen bellies. Frogs were in a frenzy of calling, their foamy nests lining the ephemeral pans, their hooting, clinking, and chiming making for a deafening chorus. Mole crickets pierced the night with their high-pitched trill as the cicadas shook up the day with incessant maracas. The brilliant carmine bee eaters filled the clay banks of the Kwando with their new nests.

In Etosha, springbok lambs lay on the new grass, so green and fresh it looked like a golf course. Lightning illuminated the pan, now pink with nesting flamingos, the waters writhing with small fishes and giant bullfrogs poised to consume their siblings. Lions lay on their green carpets, exhausted by the antics of their cubs.

During this time of renewal, termites added extensions to their colonies with the rain-softened clay, and a frenzy of winged would-be royal newlyweds took to the skies, providing a banquet for the nesting bee eaters. The few survivors scrambled along the ground, found mates, and dropped their wings, committing to breaking ground for a new colony. Edible mushrooms that tasted like the earth's ambrosia were the size of dinner plates and thick as a loaf of bread as they sprang through the sides of the termitaries.

The Linyanti swamps flooded, providing fertile soils for crops and grasses. But for those outside the flood zone, the local rainfall patterns explained why the farmers were so superstitious about their fortune or plight. The rain fell in such localized patterns that 2 inches could have fallen on one man's field and absolutely nothing on his neighbor's. Clearly, the witch doctor had carefully positioned a rain cloud over one field and not another, or perhaps on the whole village, barring one unlikable individual.

The baby elephants came among all this rebirth. In their little footy pajamas, they ran next to their mothers or after their siblings, desperate to keep up, their rubbery trunks flopping in front of them. As the wet season bloomed in its splendor, seeds were planted and crops flourished. But in the back of everyone's mind lay anxiety over the elephants. We all knew it was inevitable, just a question of when they would get the first whiff of ripe corn.

Under cover of night, elephants would tiptoe across the floodplain into the farmlands, crossing the Kwando River, roaring their prebattle cries as if they knew they were about to commit a crime or even engage in all-out warfare. Then came the all-night drumming sessions, screams, fires, shotguns, trumpets, the fury of both human and elephant, summoned up from carnal urges, beast on beast, survival of the fittest.

In the beginning, I spent many weeks going with Matthew on his trips through the region, meeting with headmen and game guards, learning the lay of the land. Our entry into this world was not easy, and it took us a while to penetrate the villages, the tribes, the politics, and the suspicions. The nongovernmental organization (NGO) that Matthew worked for, IRDNC, had been operating in the region for several years, so I was able to ride on the shirttails of their good deeds, which simplified my introduction enormously.

The last introduction Matthew scheduled was in the Mfwe area, which was just across the river from Susuwe. On the day we were supposed to meet, Tim made a plan to take our truck to Katima for a vehicle service. We woke up early, ate breakfast, and as Tim left for Katima, I looked for the least wrinkled skirt I could wear for my day in the villages. A local woman could be fined for not wearing a skirt in the village, so I wore one out of respect. I turned on the short-wave radio. A BBC reporter droned on about violence escalating between the Hutus and Tutsis. Static, crackle, white noise. I turned it off.

Matthew arrived at our barracks just after 8:00. "How's it, Caitlin?" He sat down on the porch and loaded his empty pipe. His shirt pocket was stuffed with tobacco, his brown cap rim stained with a band of perspiration. "How's the new place coming?" He carelessly lit his pipe.

I explained the delays typical of the rainy season.

"Right." He sucked. "That cottonseed soil is a bloody nuisance this time of year. Bet you'll be glad to get out of this place. Hell, it's like an oven in here."

"It's not that bad. I like the view."

"Guess I'm spoiled living at Buffalo Lodge, right next to the river. Hell, it's nice."

"Some tea?"

"Cheers, thanks, but I had some just now. We'd better get going."

I gulped down my tea, grabbed a notebook, measuring tape, and GPS, and we left the ranger camp, crossing the river toward the villages. Matthew planned to introduce me to the induna, or mayor, and a few of the game guards in the Choyi area. We first picked up Manias, the head game guard for the northern region of East Caprivi. He squeezed into the front seat next to me, smiling weakly, hesitantly, as if calculating the impact my position would have on him. Another man got in the back, barefoot, holding a pair of dress shoes to don at his destination.

When we got to the induna's kraal, we found an old man waiting on a stool just outside the small reed-enclosed compound. He was frail, dusty, clothes tattered. "Musuhili Mudella." Matthew knelt down and clapped one hand on top of the other using the reverent term for old man, *Mudella,* as the elderly man clapped weakly, mouthing his greeting in return.

"How is the induna today?"

Manias stood next to them, interpreting the Mfwe language for Matthew. "He is not feeling all right."

"Malaria?"

The old man nodded.

"Oh, hell."

He mumbled again and Manias reported, "He asks you to come back next week."

Suddenly there was a commotion on the other side of the kraal. A young boy burst out of the bush, "Papa! Papa! Leto! Leto!" He saw us and stopped suddenly, staring at me. He then looked at the old man, who nodded and opened the door to the courtyard of the kraal.

The boy disappeared inside as two wailing women streamed out from the acacia thicket in faded native wraps, each with a watermelon perched on her head. They, too, stopped as soon as they saw me. They brought the watermelons down as they stared at me, half embarrassed, half angry. I felt I was invading their privacy and tried to supplicate to them by turning my eyes downward.

When they heard me mumble the greeting "Musuhili," they burst out laughing. Rotten teeth, wrinkled faces, hands over mouths, watermelons pressed into their stomachs to prevent their sides from bursting from hilarity. They laughed and laughed, imitating how I pronounced their morning salutation. "Musuhili," they mumbled mockingly and burst out laughing again. The old man spat a barely audible reproach at them, and they composed themselves, staring at me again. He motioned for them to go inside. As soon as they were inside the kraal, they returned to their original mission of reporting the previous night's elephant mishap to the induna.

Matthew leaned over to me, "The induna's wife."

We stood there listening to the loud rapid-fire complaints from the women inside as Manias whispered an interpretation. "Leto!" they exclaimed, using the Lozi word for elephant. "The elephants have eaten everything last night. They banged drums, but the elephants chased them into the induna's brother's field. When Raymond ran out of shots, the elephants chased them again." There was another moment of silence before the door slowly creaked open, held by the little boy. He spoke to the old man, who turned to Manias and relayed the message.

"The induna would like to speak to you now."

We were led inside the courtyard, where the women helped a desiccated gray man out of his sleeping hut and onto a reed mat. He was bundled in blankets, sweating and shaking as they guided him into a sitting position, leaning up against his hut. The boy placed a long wooden bench at the other side of the reed mat. The women stepped away, sinking into the shadows. After a long silence, the induna collected himself enough to look up at Manias and point to the bench. We all knelt down and performed our greetings before taking a seat. The induna looked away from us.

"Musuhili Induna Maplancha." Matthew reached over to shake the induna's hand in between his hand-over-hand clapping. The induna weakly shook his hand, clapped, and then rebundled his blanket. "I understand you have malaria." Matthew spoke loudly.

The induna nodded.

"Do you have enough chloroquine?"

Manias interpreted and the induna spoke.

"Don't you have any better medicines?"

"I'm afraid not."

"There is too much itching with chloroquine."

"Yes, it is a problem. Induna, I came here today to introduce you to Caitlin O'Connell. I can see this is not a good time, however."

"Carry on." He waved an imperious hand at us.

"Caitlin will be working with the farmers to help them keep the elephants away from their crops."

The induna nodded.

"She will be working closely with the game guards to conduct different experiments with the farmers who are getting hassled by elephants."

Manias translated and the induna then spoke to Manias quietly for a long time, gesticulating weakly as the women interjected here and there, trying to restrain themselves. Finally, the induna's wife yelled, pointing at me, "The government must pay!"

The other woman hissed, "Nature Conservation's 'cattle'!"

I was surprised to hear English.

The induna hissed and shut them up as he continued his discussion with Manias. Finally, Manias turned and spoke to us.

"We would like her to work here with these farmers," Manias reported.

I looked to Matthew for a response.

"Yes, we are deciding where she should spend her energy, and Choyi would be a good place to start."

Manias explained and the induna nodded.

"The induna would like Caitlin to set up her experiment in his brother's field."

"As soon as we can get more supplies, we will do that." I was nervous to commit to a specific date, not knowing how long it would take for our wire order to arrive in Katima.

The induna looked pleased.

"Clatius," the induna whispered hoarsely. The boy approached, received his instructions, and then turned to Manias.

Manias explained, "The induna would like us to see the damaged field right now. Clatius will go find Raymond to show us the field. Ernest will take us there." Ernest was the induna's eldest son, the game guard for that section of Choyi.

Matthew looked at me and nodded in agreement as Clatius ran out of the kraal. Manias told us to follow him. We said our good-byes and followed Manias down a sandy path that led to a borehole where some women were washing clothing. They watched and clucked their tongues as we passed, apparently having heard the news of last night's elephant raid. We passed a cement schoolhouse painted yellow and white with graffiti written on the side saying, "Subia out," and another saying, "DTA." I asked Matthew what the graffiti meant and he slowed down to explain.

The Subia were the tribe on the eastern floodplain who held the teaching positions in the Caprivi, and this was a Mfwe area. The Mfwe didn't like the Subia, which occasionally spurred demonstrations, wherein a few people would be beaten and then the dust would settle.

"So the Subia teachers left?"

"No, not necessarily. Although one was shot in the leg last year and died of blood loss." Someone had tied the blood-soaked pants from a flagpole, and that scared people. But now the conflict had quieted down.

"And DTA?"

"DTA is the political opposition to SWAPO. Means Democratic Turnhalle Alliance, but hell if I know what they stand for."

As we caught up to Manias, we passed a long line outside the clinic. "All waiting for quinine," Matthew explained. "And some for calamine to counter the itching." Many shivering forms hovered under the blaring sun in heavy acrylic striped blankets. Some tsk-tsking emanated from the queue, and a surge of angry chatter emerged.

Manias whispered translations reluctantly. I wanted to know what I was up against.

"This year, the government must pay!" One man pointed at me, viewing me as the government incarnate.

"We should not have to work to feed Nature Conservation's cattle so that our children can starve," one of the women lamented aggressively, pointing toward the river where the elephants cross out of the reserve at night to raid crops in the village. She shifted the weight of the baby strapped to her back and retied the bundle, tightening the thin cotton patterned cloth across her shoulders. A sickly child clung to her leg with glossy eyes, having lost the reflex to brush away the flies that covered his lips.

"Chief Mamili told us that Nature Conservation will protect our fields this year, but where are they?" A man held out his hand to a neighbor, who filled it with snuff. He poured the contents into a vessel around his neck and weakly snuffed the rest into his nose. "If my neighbor's cattle entered my field, they would have to pay." He hacked from the rush of nicotine. "I should not have to build a fence to keep out someone else's cattle." He looked up and down the line for confirmation. "If the government says we are not allowed to shoot elephants, then they must keep them out of our field!"

"Yes!" agreed the woman next in line.

"Aha!" exclaimed another as the clinic opened its doors.

Matthew gave me a look, signaling that it was time to move on, and we went to the Khuka shop, where Manias had arranged to meet up with Ernest. The Khuka shops were often used as a meeting place. Ernest was just finishing a tombo, beer made from mahango, the local grain, costing 20 cents a cup.

Matthew rolled his eyes. "Morning, Ernest."

"Morning, Mafews." They all pronounced his name like that.

"How's it?" Matthew knew Ernest was not going to be much help.

At that moment, a swarm of old women surrounded me, touching, pulling, pawing, singing, dancing, stumbling with rotting tobacco-stained teeth. They slurred greetings, laughing, holding my hand, and spitting into my palm. It was intimidating, but I tried to keep from panicking and let them have their fill. They clucked, whispered, and jeered the same mocking request into my ear, wanting me to give them tea and food, holding their bellies and putting fingers together in front of their lips. I looked at Matthew for guidance, but he simply shrugged.

Manias whispered to Ernest, who nodded in agreement, pressing down the lapels of his game guard uniform. It was considered an honor to have a job that required a uniform, but the honor did not necessarily beget sobriety. We left the Khuka shop, and Ernest led us down the main road, bringing us back to the vehicle. Ernest guided us down a little-used track that meandered, becoming a footpath before disappearing into the cornfields along the floodplain.

We met up with Raymond, who told us in English what had happened the previous night with the elephants and what had been happening all season and the previous seasons. One of the rangers tried to shoot an elephant in his field last year, but he missed and shot off a piece of his tusk instead. The man lost his entire crop that year.

Walking through the sparse dried stalks of corn, we followed the path of the "Dirty Seven," the troublesome clan of bulls that had been marauding crops up and down the Kwando all last season. They always appeared together and were becoming progressively bolder. The ministry was sure that if it were to remove one of them, the rest would go away. But the plan proved to be harder to execute than expected. Orchestrating the meeting of the offending elephant and the ministry rifle in the same crop at the same time was almost impossible.

The elephant tracks were clear, coming from the reserve, then across the Kwando River, and straight through the middle of Raymond's field. Broken cornstalks accompanied the footprints as the elephants ate their way through the field before locating the pile of corn at the edge. The tracks left the farm to the south toward the rest of the village crops that sprawled in patches down the floodplain.

We measured the length of the hind footprint of one of the offenders to get a sense of the animal, as the length of the foot scales with age. At about forty years old, this one was in his prime. We looked at some other prints. The soft sand made perfectly detailed casts of their feet, even showing the patterns of fine veinlike cracks. The smaller, oblong back feet stepped into the huge round disks from the front feet, almost like craters on the moon. There were small crescent imprints at the front edges of these disks where the toenails dug in, more if running, less if walking. Sometimes distinctive cracks in the feet gave certain individuals away; other times, there could be a limp or a dragging foot, betraying an injured offender.

A lot could be read from the tracks of elephants, and the more time I spent looking at them, the more I wondered about their makeup. It made sense that a twelve to fifteen thousand pound animal would have huge feet to help distribute its weight, but the feet seemed disproportionately large compared to those of other very large mammals such as the rhino. The dense fat pad that cushioned the foot made me wonder if it was used to distribute weight or if it served some other practical purpose as well.

An elephant's fatty footpad looks like a platform shoe, adorned with cartilaginous nodes interspersed in the dense fat. A CT (computerized tomography) scan reveals that elephants really do stand on their toes, cushioned by this platform of fat. It is very dense fat, similar to the acoustic fat found in marine mammals, and, coincidentally, it was at one time prized by native cultures for candle oil, akin to whale blubber.

Acoustic fat such as the "melon" in a dolphin's forehead

serves as a sounding board to transmit acoustic vibrations into the water, facilitating sound transmission between two bodies of different densities, providing water–air impedance matching. Impedance matching is simply the matching of densities between two materials. This is needed because vibrations typically travel at different speeds depending on density—for example, the water, and the dolphin's air-filled middle ear. Sounds travel from the water environment to the dolphin's ear through fat-filled cavities of the lower jaw bones. Sounds received in the jaw are conducted to the middle ear and then the inner ear, thus eliminating the need for an ear canal and ear drum. So nature makes up for density differences by matching the impedance with fat and bones.

I later learned that elephants have this kind of fat not only in their footpads but also in their nasal cavity (like the dolphin melon) and cheeks (like the manatee). I wondered if this fat could facilitate the detection of ground vibrations by matching the difference in impedance between the ground and the elephant's foot. Indeed, pressing down on these footpads would enhance the coupling of sounds between the ground and the foot and facilitate sound transmittal to the ear via the foot bones.

After following the footprints, I began to look more closely at the fields themselves. Given the random distribution of plants in the 12-acre plot, it was hard to imagine how farmers got much out of a crop, elephants aside. There were several farmers in the region who had the wherewithal to rent a tractor plow just after the first rain for ten Namibian dollars from the Ministry of Agriculture. Their fields had comparatively impressive yields. Most farmers, however, walked behind an ox plow, doing the planting themselves, sometimes even using a toe to insert the seed. Naturally, these fields were not as flush as those where the mechanized option had been used and thus were much more difficult to evaluate in terms of elephant damage.

In those early days, I tried to estimate how much an average elephant crop raid represented in terms of bags of ground grain (the price of replacing the stolen goods). But this was a difficult proposition, considering the patchy distribution of plants in many of the fields.

Standing over a pile of corn that had been harvested the day before, we estimated that the elephants had eaten more than half of it. I suggested that leaving the harvest in one big pile made it a lot easier for elephants to stand in one place and consume the whole crop with little effort—much like a buffet. Raymond laughed and agreed; however, it made no difference to the elephants whether the corn sat in the field or in the grain storehouse. They would come to raid either place.

"But weren't elephants less likely to enter the village?"

"Yes," Raymond agreed, "less likely."

The pile had not been collected the evening before because it got too late and there was no transportation available to the women, so they decided to leave it for the night. The seven bulls had not yet been seen in the area this season, so the villagers had hoped the harvest would go undetected.

I asked Raymond if I could collect a sample of each of the grains that he was growing to compare the nutritional value of grain with preferred forage such as acacia pods, terminalia branches, and bark. He was growing three different varieties of grain: corn (a variety commonly called mealies, which could be plucked off the cob by the kernel and eaten or ground into a meal), mahango (to make the local beer or flour), and sorghum (dried and ground also into a type of flour).

After taking a small sample of each, an old woman appeared, Raymond's mother, and handed me a roasted cob to eat that she had just made in the shade of her field lookout, the place where the women spent the wet season, shooing off the birds, baboons, and warthogs by day, and elephants and hippos by night, occasionally even a neighbor's herd of cattle. But these days, many farmers were no longer sleeping in their fields (something I wanted to examine further). Traditionally, farmers migrated to small wet-season huts on the edge of their crops so that they could be there to protect the fields.

Neither elephants nor farmers were new to the area, of course, and in the past, the men had fended off elephants during the night by drumming, banging on pots, and lighting fires. Some farmers reported a greater number of elephants in the region these days, however, and most had become accustomed to the light and noise, to the point of charging the farmers and becoming dangerous. Others simply felt that the crop-raiding problem was the government's fault, since the traditional hunter who would have dealt with elephant offenders was now outlawed. The situation was further complicated by the inevitable modernization of the Caprivian household, where ownership of material goods was becoming more and more common and farmers were no longer willing to risk leaving their possessions unattended to spend the wet season protecting their crops from elephants.

Raymond's mother wanted me to know that elephants were smart and wouldn't bother eating bark when there were much better things to eat in the wet season. She watched as I plucked off a smoky kernel and ate it. It was really good. She smiled and laughed at my hesitant fingers, pulling one kernel off at a time. She pulled out another cob and brushed her thumb forcefully across the surface, creating a handful of kernels. I smiled humbly, trying to repeat her action, and thanked her for sharing her harvest with me.

Taking my mealie cob with me, I began to measure the field with a GPS in order to estimate how much wire we would need to enclose it. I wanted to use something I knew would work. The trip alarms I was developing were still experimental, so electric fencing, though a lot more expensive, was less of a risk. We promised Raymond and his mother that we would come back with fencing materials as soon as we could.

Given all the potential deterrent schemes we considered, electric fencing seemed to hold the most promise. Other suggestions had been to plant border crops that would be offensive to elephants, such as euphorbia, which produced a stinging milky sap that elephants avoided, or capsicum (red chilies), which could also serve as a cash crop. Still another barrier possibility we discussed was the type of deep narrow trenching that farmers in India dug to keep elephants at bay. Elephants avoided these ditches because they were simply too deep to climb out of once they had fallen in. If maintained, they could be successful deterrents, but that was difficult in the wet season, when the mud slid to create sloped banks where an elephant could cross. And, of course, it was very labor intensive to trench many kilometers in order to protect large communal crop areas. Electric fencing also seemed easier to erect, and it was movable and ready to set up immediately, whereas all other options required more time, money, and labor, as well as profound adjustments in agricultural practices, such as planting border crops or plant barriers around permanent fields.

On our drive home, Matthew and I made a plan to set up a length of fencing the following day at Susuwe; we needed to practice, because neither of us had ever worked with it. Little did we know, we would soon become experts.

AFTER WE'D BUILT SEVERAL electric fences, slept many fitful nights in the fields, and spent countless days trying in vain to appease angry farmers, I was still determined to help them see what was necessary to take control of their situation. But it was a slow, frustrating process. They were a tough crowd, jaded by an ill-equipped ministry and a resignation to powerlessness. There were many arguments with uncooperative farmers and frustrated local officials. An angry farmer even tried to light my truck on fire while I was installing an alarm in his field along the eastern floodplain.

After a whole year of negotiations, we finally set up an electric fence at Lianshulu village. The two-strand steel fence, strands mounted high enough for cattle and farmers to pass underneath but the spacing narrow enough that an elephant would touch the two strands with the tip of its wet trunk, was powered by a solar panel.

The locals suspected that the ministry was trying to move the boundary of the park by putting up fences around the farms rather than the park itself. But was the park really big enough on its own to host the Mudumu elephant population, which also enjoyed free reign over the region and knew which territory was safe and which was not? If the fences were smaller and just around the fields, there would be less area of fence to manage and the farmers could maintain the fences more successfully. Given a choice between fencing the wildlife and fencing the farms, we tried to convince them of the benefits of the latter. Preventive measures were difficult to sell to a people who felt that elephants were the irresponsible government's cattle.

The village had been moved out of the park twenty years prior. Many government promises had been broken, boreholes were never drilled, roads never made, clinics not delivered. The people had reasonable grounds for suspicion.

I felt powerless to help the farmers, yet also compelled to act. One of my preventive experiments would occasionally show promise, and I managed to escape the dog house, if only temporarily. These were the most rewarding times. A success meant more trust, leading to closer relationships with the women, which I especially appreciated.

Meanwhile, the elephant wars waged on, and I tried to visit as many farms as possible with my meager bag of tricks—a trip alarm here, a spotlight there—while saving the electric fences for larger areas. The solutions were life-sustaining to the farmers. Anything that I could offer was a glimmer of hope, an offer not to starve that year.

4

MAKING SENSE OF
THE ELEPHANT

*Elephants swallow a pebble every year
to keep a count of their age.*

—SHONA LEGEND

TIM AND I MANAGED TO SURVIVE the chaos of the wet seasons,
and as the elephant problems reached their frenzied height at the
peak of the harvest, we knew that a brief respite lay around the
corner. Once the last mealie cob was picked and there were no
more crops for elephants to raid, we slipped back to Etosha to
spend June and July in peace. I was always eager to get back to
the elephants, to get close to them and listen to them, observing
them daily, stepping back into their slow and contemplative
world and away from my role as crisis manager in the Caprivi.

When I had taken this position, I had no idea what I could
add to a field to which so many distinguished researchers had
devoted their careers, because I had no prior experience
studying elephants. I had so much to learn from other scientists' research, and I didn't feel like I belonged. And in a field
that had such strongly polarized political camps, I was nervous to ask for help and didn't know who to approach. I had
begun to realize, though, there was plenty more to discover

about elephant communication. This was a great source of inspiration. As with any subject, the more people who study it, in different contexts, the more there is to reveal.

I had come from studying animal communication in a way that examined concrete, measurable data, where an insect, say, might produce a certain type and number of chirps based on its genes. At first, I didn't know what to make of the communication I observed in these elephants. Despite my very different background, perspective, and area of expertise, I hoped to use my background to be able to contribute to the field of elephant research in some new way.

I pored through a collection of reprints on elephant feeding habits, ecology, movements, and behaviors kept by our project manager, Malan Lindeque, the most experienced elephant researcher in Namibia. His collection, the only resource I had available for several years, is extensive and allowed me to get up to speed on previous research.

Of course, I was particularly interested in communication studies, about which I had the least information. It was clear from my readings, however, that tactile, visual, chemical, and acoustic communication was important to elephants. I was hoping to add seismic communication to this list.

Researchers had already established that elephants produce low-frequency audio vocalizations, sounds in the range of 20 hertz or so, which is at the lower threshold of human hearing. Fossils reveal that elephants at one time had ears capable of picking up higher frequencies, but eventually they returned to the original, simpler, or more primitive, low-frequency version. Although it's not clear what kind of evolutionary pressures might have caused this shift, the impact of climatic changes on elephants' habitat, spacing, and social structure forced modern elephants to communicate over increasingly long distances, and developing ears sensitive to lower frequencies would certainly have facilitated that communication.

Hearing is a sophisticated process in mammals. In fact, along with fur and mammary glands, the three middle ear bones are a defining characteristic of mammals. Similar to the reptilian jaw, these three ossicles greatly amplify sound waves traveling to the inner ear, or the shell-like structure called the cochlea. The longer the wavelength of a sound, the farther back it reaches along a thin membrane within the cochlea; thus, ears that are more "tuned" to lower frequencies have larger cochleas, with longer membranes. The elephant and the blue whale have an enormous cochlea, allowing them to detect very low-frequency sounds. In this case, detection is through the movement of tiny, very fine hairs in the cochlea caused by the pressure wave of the sound, which is then converted into an electrical impulse and sent to the brain.

Another group of researchers had further established that elephants are able to hear sounds as low as 16 hertz, in the range of the lowest rumble in a thunder burst. They discovered this by building an elephant-sized soundproof room and training a captive elephant to touch a target with her trunk when she heard a sound from either the left or right ear. But with this experiment, they also established that it was necessary to increase the volume at lower frequencies.

This case holds true for humans as well, which is clear if you just listen to the bass portion of symphonic music. If it is to be heard, the bass portion has to be played louder. This may explain why elephants produce such loud vocalizations—so that they can be heard. As a by-product of such a high sound pressure level, the vocalizations may also be able to enter and travel through the ground.

When I started my elephant research, there was only one study describing the different types of vocalizations produced by elephants in different contexts, and no one had analyzed what features constitute these different vocalizations; that is, just what was it that made a "let's go" call a "let's go" call? Was it simply context or was there something inherently encoded in the call that defined it as "let's go"?

As I was collecting every call I could in every context imaginable, it wasn't immediately obvious what the differences were, if any. I was able to distinguish between aggressive or excited calls and those made under more relaxed conditions. The frequency modulation was much greater in the aggressive or excited calls such that the pitch of the call would start low, go higher, and then return back to low pitch, as opposed to a very monotonic contact, or "let's go" call, which had very little rise and fall in frequency. And these calls tended to overlap more than others.

The initial "let's go" rumble elicited specific vocal responses from the herd, followed by a perfect choreographed exit en masse, but the responses to this call seemed indistinguishable from the first "let's go" rumble to which the elephants were supposed to be responding. I was determined to figure out how the elephants were interpreting these messages. Was it the act of stepping away from the waterhole and keeping still that tipped off the other cows that the matriarch was ready to leave rather than the specific rumble that she had made? None of the others seemed to look up from its drinking to be able to notice this. Was their leaving triggered by a combination of visual cues, context, and the vocalization? It didn't appear that visual cues were that important in the initial stages of leaving; nor did their leaving seem to be solely situational. Perhaps elephants could recognize individual voices, just like we do, but measuring what defined these differences required a level of software sophistication that I hadn't had access to in my early studies, so I tried to focus on what I was able to do with available technology.

Here in Etosha, I began to explore ideas to enhance my rudimentary elephant trip alarm to discourage crop-raiding elephants in the Caprivi. My goal was to align the trip alarm design with the elephant's own communication system. My initial design was pretty low-tech: car alarms, clothespins, tacks, nails, and string were the principal materials with which I experimented, and the results were promising, but I worried about the long-term success of the technique.

At the time, the farmers were very pleased when my system worked; it either scared the elephants, or at the very least, woke up the farmers. If the farmer was wealthy enough to own a shotgun, the noise of the shots reinforced the threat of the alarm. But I worried that the elephants would get used to the sound of the alarm and realize that there was no immediate threat. I wanted to design something that might have more clear meaning to the elephant and thus perhaps take longer for the elephant to habituate to it.

I thought that the elephant's own alarm calls might make for a more effective tool to chase them out of the Caprivi farms. The car sirens worked for a season or two, depending on how many times they were set off, but the elephants grew accustomed to the noise over time. Success also depended on how large an area needed to be protected. A 550-yard radius around the alarm was about the limit of its effectiveness.

In contrast, a call that actually had meaning for elephants might not only be a greater deterrent but might work over a longer distance and for a longer period of time. I understood that the low-frequency vocalizations in the ground had the potential to travel farther than airborne signals. Perhaps an alarm system based on seismically transmitted warning calls could be more effective than my original makeshift design.

I began by analyzing the alarm call sequence I had recorded the previous year at Mushara. The calls had been made to warn of hunting lions, and three of them contained much more rise and fall than any of the other calls I had recorded, even those made during other aggressive interactions such as with rhinos. I prepared a tape to play back to different herds in three regions of the park to see if elephants from other areas would respond similarly to the herds at Mushara and to determine which portion of the call sequence had the most visible effect on their behavior. To play back the tape, I used nothing more sophisticated than a boom box, since it was something that a farmer might own.

Armed with my new alarm call sequence, I returned to Mushara, eager to try it out. I conducted all the playback experiments from the hide, not knowing whether elephants would respond aggressively to the deep rumbles emanating from my pickup truck. Before each playback, I gave the elephants a chance to drink while I took careful notes on the group's time of arrival, size, age composition, and distinguishing features. Of course, I recognized some members of the herd, such as Broken Ear, Left Tusker, and Collar. I also filmed each encounter.

I played the tape to six different family groups, and each time I saw the same response: heads up, ears out, and then tails straight out behind them as the entire herd ran off in silent terror. There was no "let's go" rumble, no lining up, just a speedy mass exodus. Sometimes it took more than a day for any one group to feel comfortable returning to the site.

I was shocked that the calls would elicit such an extreme response, so I was careful not to repeat the experiment too often before thinking more about the next experimental design. I did not want to torment the elephants unnecessarily. I pored over the videotapes to see if I could figure out exactly what had happened, to determine who, if anyone, had made the decision to leave and if there was a particular part of the call that was more effective than the rest. I also noticed an occasional delay in response to the calls and wanted to investigate that further.

While waiting for the family groups to show up, I also tried the experiment with bulls. Without exception, all the bulls had a very similar but less reactive response. They acknowledged the calls, shook their heads aggressively, even stepping away from the waterhole for a moment, but then returned and resumed their drinking. Bulls probably were not as affected by these calls because they do not have the same fear of lions as a breeding herd, where the females have to remain constantly vigilant to protect their young. Warning calls would clearly have no bearing on crop-raiding bulls. Since most of the crop damage in the Caprivi was caused by family groups, though, this discovery didn't discourage me.

Back at Etosha Ecological Institute, I analyzed the videotapes and discussed the results with Tim and other colleagues. The footage amazed everyone. No one could believe that elephants would react so strongly to a simple recording. Three rumbles in particular out of a series of low calls, growls, and trumpets appeared to be the source of the dramatic response. So feeding the original digital recording into the sound analysis software on my laptop, I cut and pasted these three calls into a separate recording and made a new cassette tape containing just these.

I returned to Mushara to play the new tape and see if the recording would induce the expected response. Indeed, the result was the same, only the response was more immediate, as the first several calls on the original recording did not have an effect. These first calls had served as a control so that I could be sure that the elephants were not responding to my presence or to any of the noises I made, or even to their own calls that were not specifically the intense alarm calls that triggered a flight response.

On my last night at the site, Tim was able to join me to witness the elephant responses firsthand. They were as impressive as before, and I appreciated having a second person there to experience it. The following day, Tim helped pack up, but I was so comfortable with this site that I no longer felt the need for company to set up or break down. It was just nice to share the space with him.

At other sites, however, I hadn't become fully comfortable. Some still made me a little nervous. Toward the end of my field season the year before, I had asked Tim if he would spend the night with me setting up a new site in the bunker at Olifantsbad waterhole, a veritable elephant amphitheater. It was different from Mushara because it was not as far from other water sources; therefore, the elephants traveled in smaller groups of fifteen to twenty individuals rather than within bond groups of between thirty and fifty. This smaller scale made it easier to study them and make recordings of family groups. Here, the herds choreographed their arrivals and departures: if their timing was off and they arrived too early, one herd would stand off at the edge of the clearing to wait its turn. The interactions between individuals were also richer at this site, less stressed.

I was just beginning to understand the relationships within and among family groups, and I wanted to spend more time focusing on how matriarchs made decisions about how to share resources such as water, how they made decisions to leave the waterhole, and how long it took them to execute their plan. These questions were more difficult to consider at Mushara because the interactions between large bond groups were often more chaotic. Arrivals at the waterhole could involve more than a hundred individuals, usually with an upset here, a spat there. It made sorting out family groups and making decisions about them virtually impossible at times.

At Olifantsbad, I was particularly interested in why Bent Ear was no longer the matriarch of her family and when exactly the herd decided that her daughter would take over the role, if in fact it was her daughter that had assumed the lead. Bent Ear was a very old cow, a little slow, and had a distant wildness in her eyes not unlike that of the elderly insane, but I had never read anywhere that "grandmothers" retired their matriarchal role or how such a retirement would come about. It did seem reasonable, though, that there would be a system wherein the matriarch would pass on her role to her daughters, or, as other colleagues have shown, to the next oldest herd member, prior to death.

I wanted to observe what role the retired matriarch had in the family, if any, and had spent a few nights in the back of my truck, in order to scout out the site. But the lions were getting so bold that they climbed up on the bumper to smell me inside my tarp-covered pickup. The sinking weight of their paws forced me to retreat to the bunker.

After setting up my recording equipment next to the hide, I built a rock cairn over the microphone to protect it from the insatiable curiosity of young bulls. I ran the wires through the slit in the hide and down to the tape recorder next to my head. That way, I did not have to get up in the cold to record, unless elephants arrived; then I would bring the recorder up to the bench and record while observing with the low-light binoculars.

Just before sunset, Bent Ear arrived, bringing up the rear of her herd. I sat on the bench, enjoying my new vantage point, taking notes on the elephants' arrival time, notable interactions such as who got to drink at the outflow, whether anyone got displaced in the process, and whether any allomothering was apparent, where a sister or aunt would watch over a baby while a mother drank. I noted the number of vocalizations that occurred throughout the visit. I noticed that the vocalizations increased in number after the matriarch's first "let's go" rumble.

Being so close to the elephants, I was able to get some fantastic photographs, which would help me identify them in the future. Bent Ear's family group did not stay long because another group waited on the outskirts. The rumbles volleyed back and forth until Bent Ear's replacement finally gave the call to leave. The waiting herd was led by No Tusker, and as the members entered, a young bull from each herd rushed forward to greet each other. They wrapped their trunks in an elephant handshake, placing the tip in the mouth of the other. Once the greeting was over, they pushed on each other with the base of their trunks.

Suddenly one of the young bulls realized that his herd was almost to the edge of the clearing and ran off wailing after his departing family. The other approached my microphone on his way to the water, sniffing at it through the cairn. I sat silent and motionless while he inspected by placing his foot on top of it, but when he started to put his weight on it, I clapped my hands, startling him to make a hasty retreat to the other side of the waterhole a safe distance away.

One of the calves found a scratching post, an old leadwood stump next to the water. He mounted it and scratched his belly back and forth with sheer ecstasy. He had good control of his trunk for his age but still swung it around over his head in delight. Soon his older brother or cousin wanted to share the scratching post, and they ended up sparring, the older one chasing him into the water, where they rolled around with the other calves.

As the sun began to set, No Tusker eventually positioned herself and gave the "let's go" rumble twice to rally her troops. The rest of the herd lined up behind her, and she waited until the group was complete before slowly making her move. Just before reaching the edge of the clearing, No Tusker and three other adult cows froze midstride, leaned forward, one or two lifting a front leg off the ground. They took a few steps and then froze again, the rest of the herd following suit. No Tusker held her trunk up and sniffed at the air toward the west. After a few more stops and starts, they disappeared into the bush. It was now dark.

Shortly thereafter, Tim arrived. It had been ten minutes since the herd had left, and I wondered if the herd's tentative departure had anything to do with the sound of his truck. I kept these kinds of observations as separate entries in my field notebook, slowly building a database of freezing behaviors and potential sources of vibrations that the elephants could be sensing. I suspected that elephant vocalizations were traveling in the ground, and this abrupt stopping behavior and leaning on toes were means to detect them.

I was happy to see Tim and heated up his favorite bush delicacy: a can of mutton curry. We sat on the bench together looking out at the waterhole, he with his mutton and I with vegetable curry. It was a new moon, but the calcrete stones glowed as brightly as the stars. It was dead silent except for an occasional blacksmith plover peeping a high-pitched objection to a passerby. We sat for a long time, enjoying the peace and the occasional shooting star. A distant chorus of jackals lit up the air and then petered out.

The night turned out to be less peaceful than expected, since we discovered in the middle of the night that the Olifantsbad pride had youngsters whose shoulders could fit within the slit of the bunker. I had heard roars a few kilometers away. There was silence again for so long that I dozed off, only to be startled by a clink just above my head. The sound was coming from the calcrete rocks piled up around the hide to prevent lions from climbing up. I waited for more footsteps. Nothing. It was such a soft sound that I thought it must have been a jackal. I lifted my head to look through the slit and saw a head peering in at us. Not a jackal but a lioness. I whispered obscenities, hoping to startle her away. She sat crouched down, taking up the whole slit, undeterred by my weak pleas. I woke Tim, who immediately leapt up, incredulous at my calm, and shined a flashlight at her. The intruder had her shoulders inside the hide, paws extended, searching for ground. If she moved forward any farther, she would have fallen in.

Tim roared, banging the equipment trunk with his dinner pot, while I rummaged through the trunk and attached the alarm. The sound pierced the night as we sat there undressed, frigid, flashlights stabbing at the interloper. She bared her teeth. My heart leapt into my mouth. Tim hurled himself at her, armed with maglight, bellowing curses from some primal place inside.

The lioness reluctantly shifted her shoulders, and just before the maglight hit her in the nose, she calmly removed herself, then slowly padded off to the left of the hide. Our beams followed her, suddenly reflecting off a sea of golden saucers, blinking in objection to our lights. Gold means predator; red means prey. The pride was waiting below.

Livid, Tim started pulling rocks through the slit from the inside, stacking them into the hole. I slowly followed suit. He insisted that we not stay. The gaping hole in the roof and the door that would not close finally convinced me. Reluctantly, I packed up my things, consoling myself with the thought that one of the lion researchers might be willing to help me make some improvements to the hide in the future. The pride watched with indifference at the far edge of the pan as we roared off, two trucks in a dusty, late-night convoy.

Having spent a few days at Okaukuejo to regroup and make a new plan, we were confronted with several more lion incidents at the floodlit waterhole of the main camp. One night, in the bitter cold air at 2:00 A.M., the whole camp woke to what sounded like the terrifying cries of a human child. We all rushed out to the waterhole to find a huge black rhino calf pinned down on its back with an exhausted male lion holding its mouth shut, a female at its throat trying to crush its windpipe while another male gnawed at its genitals. With such a thick-skinned animal, the only way in is through the soft genital area. The lions were all wet, so the struggle must have begun in the water. The mother apparently ran off after unsuccessful attempts to save her two-year-old calf, barely distinguishable from his mother in size and known to be healthy. It took hours before they were able to silence the horrific cries and suffocate the terrified beast. It took several days to finish the carcass.

Shortly after this event, an old lioness came into the area and provided days of entertainment to guests at the waterhole by her many unsuccessful attempts to make a kill. The entertainment turned to horror when she finally got desperate enough to crawl under the fence of the main camp, and in her prowl, found a tourist sleeping under a tree. She and a younger male companion proceeded to dine on this poor unsuspecting person, who apparently died instantly as his friend woke in terror and ran to contact the park vet, who came and dispatched the man-eaters.

Seeing that Okaukuejo was not the best place to recover from my own lion encounter, I rallied the courage to return to

Olifantsbad to repeat the playback experiments from my truck during the day. Regrettably, I was not able to finish my observations on the rise and fall of matriarchs at this site, as it was too unsafe and impractical. Fortunately, I was able to take this question up again at Mushara later on, when I was able to build a history of relationships over time and got to understand not only how dominance developed and diminished within a herd, but that there was a clear dominance hierarchy between herds, which explained some of the waiting behavior I had witnessed at Olifantsbad in these initial studies. I realized later that the waiting was not polite, by any means, but was necessary for survival, as a matriarch defending the waterhole from unrelated herds is not a pretty sight.

AFTER NOT GETTING nearly as dramatic a response from herds at Olifantsbad as at Mushara, I chose a third site to repeat the experiments and spent several days in the southeast of the park, at Gobaub tower. Elephants from other regions of the park perceived these calls as an alarm, but didn't respond as urgently as at Mushara, acting as if the warnings were meant for someone else. They listened intently, got nervous, and eventually left, although not nearly as precipitously as the Mushara herds. In fact, the Gobaub elephants even responded aggressively toward the source of the calls by surrounding the tower where I was stationed, ears held out, making a long chain of outraged elephants, bellowing rumbled insults that eventually turned into quiet curiosity. So rather than flight, they chose to fight.

At one point during this aggressive standoff, a young bull got confused and started wailing and trumpeting until the matriarch and adult cows decided that they had made the appropriate aggressive gesture, shook their ears at me, and moved on. I was happy to be in a tower made of railway track steel. The French ranger from the Belgian Congo who built it knew what he was doing. But I became concerned for the farmers because they would not have the protection of a steel shell when they tried to use this call to chase crop-raiding elephants. Was this really going to be an appropriate tool for them?

Although I needed more trials and better acoustic control, these studies suggested that elephants may be able to identify a caller and respond accordingly. Theories of appropriate responses are well documented, since several researchers have had success attracting bulls and eliciting appropriate responses by playing estrus calls. But nothing had been published at that point to demonstrate that elephants could discriminate between callers.

I eventually included these data in a paper on the efficacy of the different elephant crop-raiding deterrents with which we had experimented in the Caprivi. Lots of questions remain as to how elephants identify individuals, as well as what constitutes a particular type of call.

The following year, another elephant researcher, Karen McComb, and her colleagues reported a similar finding, confirmed through a very detailed six-year study in Amboseli National Park in Kenya. She found that matriarchs could distinguish calls made by their own extended bond groups from calls made outside their bond groups. This research also established that some matriarchs were more confident than others and that less confident matriarchs were wary of strangers. Wariness was measured by how closely individuals were spaced within the herds (known as bunching) while listening to the playbacks of contact calls broadcast by researchers. I planned to incorporate this measure of wariness in my future playback studies because I suspected the same thing was happening with the matriarchs in my study area.

It remained to be seen, however, whether a call could be broken down into parts that an elephant would recognize: Is there a universal elephant vocabulary? What defined an alarm call and where was the identity of the individual elephant encoded in it? Could I play back different frequencies of the warning call and get a similar response from all elephants? If there was some way of creating a universal warning call, could farmers use this as a potential crop-raiding deterrent that could be transmitted either through the air or through the ground? Would it be too dangerous for a farmer to use? And, if not, how long would it take for elephants to get used to alarm calls and realize that they were fake—the boy crying "wolf"?

Those answers would have to wait. With another season drawing to a close, I decided to return to my favorite site where I savored the last few nights back at Mushara.

5

TIPTOEING
ELEPHANTS

When an elephant happens to meet a man in the desert,
merely wandering about, the animal shows both mercy
and kindness to him and even points the way out. But,
the very same animal, if it sees the traces of a man
before it meets the man himself, trembles in every limb
for fear of an ambush, stops short, scents the wind,
looks around and snorts aloud with rage.

—GAIUS PLINIUS SECUNDUS (23–79 B.C.E.)

IT WAS AFTER TWILIGHT ON A NEW MOON night when I saw three bulls on the northern horizon just before closing into the bunker for the night. I sat on top and waited for them to walk in. In the near darkness, I felt like I was suddenly in the depths of an open ocean, sitting within my little submersible, phosphorescent stars suspended in the distance as the bulls approached. Their gait was so soft and fluid that they seemed to float in the luminescent sea like blue whales in a bottomless expanse, the major and minor Magellanic clouds in the Milky Way looking like the spouting of water through elephantine blowholes in the deep.

A soft padding of heels flowed by me as the bulls walked past the bunker and on to the waterhole. Just 10 feet away, three pairs of eyes remained forward, not even a sideways glance toward me, though I knew that they were aware of my presence by the position of their trunks, just the tip turned toward me, absorbing my scent as they passed. Their breathing was slow and methodical through the end of lanky trunks, held just inches above the ground. With one exhalation after another, they calmly took me in as if everything were in order; I had finally become part of the scenery to them.

The more I immersed myself into their world, the more questions I had, even down to the simple positioning of their feet while walking. Over time, I saw that an elephant's mood could be determined by its footfalls. If it was calm, it placed its weight on its heels like these bulls were doing. But if it was upset or nervous and felt the need to be stealthy, it placed its weight on its toes and literally tiptoed.

I noticed this behavior on our first foot patrol in the Caprivi, when Tim and I stumbled upon a huge family group of elephants and found ourselves surrounded. This is one of the most dangerous scenarios in the bush. We had joined Jo and the rangers on one of these foot patrols in order to learn the lay of the land within the reserve where we were stationed. We had driven down the dusty Guesha track, parked near the pan, and started out.

A small herd of zebras galloped off into the dappled forest as we approached, leaving behind the echo of their unlikely high-pitched call, "Aoa, aoa." A large herd of buffalo grazed off in the distance, but after one sharp bellow from the sentry, the herd was off at a clip. After walking for a few miles toward the river, the Baikea dissolved into a mixed acacia forest, large erioloba trees creating heavy canopies of shade. A lone giraffe loped away.

We continued walking through the acacias until we saw a small herd of elephants taking shelter under the shade of a grove of trees. They were fanned out in a circle, some dozing on their feet, some lying on the ground within the circle for protection. It was the first time I had seen elephants lying down. There were two other herds, one resting to the left, another to the right.

Suddenly we realized that we were surrounded by elephants. How this could happen without realizing it seems difficult to imagine, but it's actually a common situation in the bush, particularly when elephants are at rest. Family groups tend to gather under the shade of a few acacias, sentries facing out, dozers inward, and often the matriarch will walk between family groups that spread out in the adjacent shady patches, quietly on patrol while the others doze. During these times, elephants are extremely quiet. There isn't even a snapping of a tree branch to give them away. If these encounters go awry for some reason, they can be dangerous, because defenses are up and assessments are made on the spot.

A large cow walked casually and unwittingly in our direction. This was bad. How could we warn her of our presence without startling her? There were a few extremely tense moments as Jo signaled for us to get down into the sparse bush. Coming upon us, she could either charge or flee. Jo pulled out his meager 9mm just in case. A 9-millimeter bullet would just make her angry, but perhaps the noise of the shot in the air could trigger a retreat. In any case, his confidence in the situation was momentarily encouraging.

The elephant was almost on top of us when she finally caught our scent. A huge gray mass loomed above, frozen in midstride, tipping forward as if on tiptoes, as we huddled under the imaginary cover of the brush. She waved her upturned trunk high in the air, her wide eyes darting back and forth, trying to see the source of the offending smell. Fight or flight? Flight!

She spun around, tail erect, mouth agape, and eyes wild, while several hundred of her extended bond group erupted from beneath the surrounding shade and vanished into the thick forest without even a snap of a twig. We heard one very soft rumble, like the sound of a distant truck changing gears, but that was it. Gone, with barely a sound.

How could the weight of that many elephants be so inconceivably silent? That they could run through the woodland over fallen trees, brush, and leaf litter without making noise was so remarkable, we simply stood in awe.

I marveled at the event for weeks. They appeared to be on their tiptoes as they ran. But do elephants really tiptoe? It seemed an absurd proposition for such a huge beast. With some further observation of elephants in stressful situations, I discovered that they do indeed tiptoe when startled or wary. But they're quite adept at getting up onto their toes for lots of different reasons, like reaching high into an erioloba tree to eat pods. On one occasion, I had even seen an elephant stand on hind legs to reach these delectables, though usually bulls will resort to shaking the tree to get the pods to fall.

One morning as I sat on our porch having tea, an old bull approached the large erioloba at the edge of our drive. The tree was full of pods in very high branches, so the bull leaned his forehead up against the trunk, locked his shoulders, and pushed the weight of his body forward and back against the tree. The bottom half of the tree stood solid as the top shook back and forth, showering pods to the ground. A younger bull waiting in the distance approached to share the spoils but was discouraged with a shake of the head. The old bull stood happily crunching the fuzzy, leathery gray pods. When he had his fill, he decided to investigate me. I sat still as he approached with his trunk extended to take me in. He stood about 3 yards away, stopped, and stared, his vaudeville eyelashes a stark contrast to his hulking presence. Finally satisfied, he released his enormous member and peed about 3 gallons onto the sandy drive before moving off into the forest, flat-footed and confident.

I looked at his giant stethoscope feet as he walked away, one following the other as he disappeared. I was reminded of the peculiar behavior of elephants investigating a dead loved one. Placing their front foot just over the surface of the skin, they scan it like a sensor over the body. Then they turn around and do the same with their hind feet. Could they be sensing for vital signs? Sensing for vibrations, a heartbeat, or breathing?

I went out to the tree, leaned against it with all my weight, and tried to imagine how strong I'd have to be to shake it as the bull had just done. I pushed my hands against the rough bark in vain. The leaves didn't even shiver.

I tiptoed back to the house, imaging myself in platform shoes, rolling my weight forward for the tiptoeing and then back for the more comfortable walking gait. In several instances in Etosha, elephants hadn't seemed concerned about a silent retreat and would run all out to approach the waterhole, with weight more at their heels than their toes. Even then, the soft pad of the heel dispersed the weight so that elephants running never really sounded as loud as you would expect. The leathery swishing sound of their legs seemed louder than the sound of their feet hitting the ground, unlike the noise of hooves hitting the ground from animals such as zebra.

One other time I had startled an elephant by accident and caused it to tiptoe spontaneously. It was just outside my bedroom window in the dripping early morning of the wet season a couple years before, after spending a tumultuous night suffering through a series of nightmares brought on by mefloquine.

After arranging the mosquito net over the mattress, I lay down and tried to read. But a flurry of insects in the room were vying for the limelight. A preying mantis threw itself mercilessly at the flame. Its wings were totally singed. I caught it gently in my hands and took it outside the room, closing the door behind me. It was doomed, but I couldn't bear to watch it die.

When I got back into bed and started to read again, a thick brown noctuid moth was at the helm. Shooting the moon with wings of fire, its desperate kamikaze mission eventually snuffed out the candle. With its remains left quivering in the hot wax, I lay in the dark, drifting off to the smell of nuts and greasy smoke.

Eskimos were playing lacrosse on the snowy airstrip as I tried to land. The updraft rocked the wings of my Cessna as I struggled to maintain a landing course. The sun was just about to rise above the earth, wedging a glowing peel next to the starry night. My right wing tore loose and the plane went spiraling down, but slowly, suspended like a weightless leaf in autumn. I reached out and grabbed the broken wing and held it tightly as I tumbled down.

I asked the two men in the control tower if they could tell me where I was supposed to go next. I knew the moon was just a stopover. Where I was going was much, much farther. But I was suffering. I had not slept in two days and I felt dangerous again. I was cold. So cold.

I WOKE UP holding my knees to my chest, shivering, thinking that I might have malaria. I looked out the window at the blue night. The moon was out, casting the forest and floodplain in a silvery sheen; remnants of rain dripped off the vegetation. Intermittent sounds accompanied the peaceful scenery: a lone hyena probing the night with an inquisitive octave, a Scops owl calling in the distance, and a mole cricket's high-pitched trill from its burrow just under my window. The crack of an elephant breaking branches off the erioloba trees near the side of the house was the comfort I needed to fall back to sleep.

I awoke for good at dawn, delicately extracting the shards of yet another nightmare lingering in my psyche. I was getting better and better at this process, but it required more time than I often had. The noise of feeding elephants grew as they moved closer. I could have reached out and touched a young cow right from where I was lying if it weren't for the screen window between my bed and that thick skin. Her lashes were almost comical in length and thickness, eyes contented as she chewed an acacia branch. I looked beyond her to see that it was a herd of about fifteen. They were very quiet except for the rustling of leaves and snapping of twigs. I lay there watching them, afraid to move for fear of scaring them off.

I watched her eyes, searching for an expression, but she was unafraid, going about her business, chewing, chewing, pulling new branches with her trunk and inserting them into her mouth, more chewing. I soothed myself with her presence, confidence making steel bars out of the thin mesh between us. I kept still. Elephants don't like to be surprised.

I shifted to get a better look at her feet, and the book that I had fallen asleep reading dropped to the floor. Immediately I saw the whites of her eyes as her pupils contracted and she tip-toed off, tail out and mouth wide open as if she had just seen a ghost.

I planned to pay much more close attention to an elephant's gait after these early observations. And sure enough, when I conducted alarm call playback experiments, I also saw the tip-toeing posture assumed just before family groups at Mushara ran off in a panic. The matriarch would cautiously advance to investigate the sound and then start to tiptoe in surprise before she instantaneously and seemingly silently rallied the troops to hightail it out of there, and literally with tails held high, but by that point, in an all-out stiff-bodied elephant run, it was all heel to toe.

I was especially curious about the sneaky tiptoeing, which led me to start thinking about tiptoeing episodes that correlated with seismic events. Could elephants lean forward on their toes to detect signals in the ground? The idea sounded ridiculous, but I was beginning to think that elephants might be using the fatty pads in their feet to detect vibrations, or help conduct vibrations via bone conduction from their toenails to their ear bones. I knew that other species like the blind mole rat had two possible ways to detect vibrations. Why couldn't this be true for the elephant?

Perhaps elephants were getting up onto their toes to detect vibrations through special vibration-sensitive cells at the tips of the toes or through their toenails. Alternatively, the vibrations could pass through their foot pads and toe bones, feet, then up the leg, and into the ear. I had seen elephant bulls at Mushara shift their weight onto the tips of their toes while standing on top of the water flow pump when the water level was low. I had taken photos of them tipping their foot forward so that their toenails were touching the cement lid to the pump. Were they doing this to feel if the pump was on? I was excited to think that there could be more than one answer, more than one mechanism to explain what I believed—elephants could detect and use vibrations to gather information.

Meanwhile, back at Mushara, I descended into my bunker submersible for the night as the three bulls drank quietly for some time. It was so dark now that all I could see were three giant black silhouettes, slowly sucking up 3 gallons at a time and pouring it from trunk to mouth with head held up. Afterward, there was a slow exhale and trickle of water, and an occasional passing of gas or blowing of bubbles in the trough with their trunk. The hypnotic process repeated itself again and again.

I watched their cushioned feet lean forward and back intermittently as they drank until I got too cold to watch from my comfortable eye-level vantage of chair on top of table. I tucked into my bedroll for the night and took notes on my tiptoeing elephant friends before dozing off to the comings and goings of more elephant bulls throughout the night.

6

THE DRY SEASON

*When the river dries up and there is nothing left for
the animals to drink, the elephant will dig a great
hole, creating a waterhole for all, thus saving
the thirsty animals from drought.*

—MAYEYE LEGEND

THE THICK AIR WAS BLOATED, sliced by the rhythmic purring
of a sickly ceiling fan. The bleak dry season could make or break
the human spirit, oscillating between tumult and calm, paranoia
and tranquility, all brought on by the wind, smoke, fires, and
ash. Heat and pestilence can do strange things, whittling away
at your sanity and giving way to delusions.

At one point, it was so hot that a rare bottle of New Year's
champagne burst in the cupboard, the sweet sap bubbling into
the parched wooden shelf. Africa is subject to this indescribably
extreme humid drought, until the clouds break, and the desiccated earth, the thirsty animals, and the starving plants soak up
the rain like sponges.

In 1993, in the midst of this, Tim and I finished our house.
On top of a previously laid 26 by 20-foot cement slab left
behind and unscathed by the army, we had a waist-high brick
wall constructed, finished off at the top with tightly threaded
reeds. With screens across most of the front and back, the open

floor plan and walls gave the feeling of being completely open to the elements—comfortable yet vulnerable. I installed rubberized window shades, fitting them with metal rods to cover the screens at the front of the house to stop the rain that would drive in sideways off the floodplain.

Tim plumbed the house, adding a flushing toilet and a shower, with a small hot-water heater—a great luxury. We installed basic wiring so that the lights, radio, computer, and printer could run off a solar panel. Later, we purchased some kiaat wood, a local red hardwood, and built a bar for the kitchen, a desk, bookshelves, and shelves for our clothing, all constructed with just a hand planer and hand drill.

Unfortunately, the house was topped with a sloppy thatch job, as we couldn't afford a good foreman or a professional thatcher. Over time, the roof started to lean and we had to moor it to a nearby tree. Nevertheless, we loved our humble abode.

If we had to be away for any length of time, the baboons would make a playground of our roof, so we had to ask Mtombo, the Bushman laborer at the station, to watch the house whenever we were away for a week or more. Baboons love sliding down thatched roofs, and with every ride, they took handfuls of thatch with them. A thatched roof could disappear very quickly under such pressure. We even covered the thatch with chicken wire, but we could only afford the medium gauge, which slowed but did not prevent its inevitable destruction.

We spent the dry seasons counting and tracking elephants, putting up electrical fences and trying to remain cool and hydrated. We conducted dung and vegetation transects, a technique of sampling an area by choosing a plot of land and observing it many times to see how it's used by wildlife. Over time, these observations were measured, which helped us to understand how elephant population density affects an area of vegetation.

In the Caprivi, we studied an open, or free-ranging, population, so the rules are different than they would be in a park like

Kruger National Park in South Africa, which had more defined boundaries. Elephant foraging had a different impact on trees, depending on the soil type. Trees in clay soils couldn't recover as well as those in sandy soils. Three elephants per square mile is thought to be a healthy density depending on the habitat, with a population growth rate of 3 percent per year. Up until the mid-1990's, Kruger managers have harvested about seven percent of the population every year, and as managers were concerned that the vegetation couldn't sustain the pressure from a rapidly growing elephant population. Such are the harsh considerations of managing elephants. In these areas, administrators must take into account all species, not just the elephant. And with such an intelligent animal at stake, the issue is contentious to say the least. After this time, there has been much research dedicated to alternative solutions.

At one time, the slow and thoughtful elephant did the flora and fauna a great service with its eating habits and inefficient metabolism; having to chew through some 500 pounds of vegetation a day in order to extract about 40 percent of the nutrients for itself. The rest was excreted, providing sustenance for insects, birds, and small mammals. Elephants often pushed over whole trees to reach choice, new growth or pods in the canopy. This habit was an important ecological force in forested habitats like the Congo basin where large mammals depended on the elephant to open up patches of forest, promoting the growth of grass and secondary growth forest and facilitating species diversity.

The plants, too, benefited from the fantastic seed dispersal and priming of seeds provided by the elephant's digestive tract. Not a picky eater, elephants enjoy more than ninety different tree species. Over time, however, it became a question of their having to rely on eating plants that were restricted to smaller and smaller patches, which some park managers felt could no longer tolerate the pressure of this indiscriminant marauder of the habitat.

Elephants were not designed for confinement; their territories have been known to extend over 850 square miles. For a species with a life expectancy of sixty years or more, usually limited by the wearing away of their molars, that's a lot of food to consume and a lot of territory to wander. Had elephants evolved with a more efficient digestive tract or foraged with a little less abandon, perhaps their management wouldn't be so problematic. Unfortunately for large, complex, long-lived animals such as elephants, human influences usually occur faster than the rate of evolution, and elephant habitat is diminishing at an astonishing rate.

Elephants in the genus *Elephus* (modern Asian elephants, of which there are four living species) were once widespread throughout Africa, flourishing there for 3 million years. But in the Pleistocene era (1.8 million to 10,000 years ago), habitat for *Elephus* gave way to the more open and drier savannah habitats preferred by its contemporary, in the genus *Loxodonta* (African elephants, of which there are three species). At this time, the genus *Elephus* crossed the land bridge to Eurasia and then eventually became extinct in Africa about 35,000 years ago. Meanwhile, the *Loxodonta* thrived in sub-Sahara Africa. It is difficult to imagine that the closest living relatives of the modern elephant are the hyrax (rock dassie, about the size of a guinea pig), the dugong, the manatee, and the golden mole of the Namib Desert.

Perhaps one day the fossil record will reveal new evidence to shape a different interpretation of the elephant's origin, but it's clear that the elephant is highly adaptable and travels great distances. But these adaptations take time. How can a species born to migrate, compost, and disperse be expected to adapt quickly enough to survive a rapidly changing sociopolitical climate? Some believe that the elephant doesn't negatively impact the environment. How can competing points of view be reconciled to come up with a reasonable and responsible management strategy?

Humans and elephants share common traits: neither appears equipped to compromise; both are refugees of war, struggling for a foothold, a patch to resettle, to reclaim and call their own. Unfortunately, elephants are victims of human circumstance; humans have more choices.

Through experiments with birth control, managers in Kruger National Park have attempted to find alternative solutions to reduce what they believe to be the negative effect of elephants on the vegetation. One method placed time-release estrogen implants in the ears of a number of adult cows throughout the park. This experiment took place in the mid-90s. Some reported that the estrogen caused an increased interest in the cows by bulls, resulting in constant harassment of these poor cows. Since it was expensive and yielded mixed results, that experiment was discontinued.

In 1996, another experiment used a protein immunization technique developed for horses, in which a protein extracted from pigs was injected into females, causing their bodies to create antibodies against this protein, which in turn blocked fertilization of the egg, thus simulating pregnancy. This technique showed great promise but was also very expensive. In 1998, it was estimated that at least three thousand elephants or more would have to be immunized every year in order for this technique to work. The research team that developed the protocol set about to develop an artificial vaccine, as it would be too difficult and costly to extract the protein from pigs to make this program viable for large numbers of elephants.

The forty-one elephants involved in the initial experiment were fitted with radio collars in order to track, dart, and take hormone samples from them annually, as well as given an ultrasound exam to ensure that the twenty-one cows given the vaccine weren't pregnant. I was able to join the team on one of these ventures because of my collaboration with Thomas Hildebrandt of Irwin Berlin on some elephant anatomy studies. An expert at performing ultrasound on elephants, Hildebrandt and his team perform them regularly in the field. Although none of the cows were pregnant, I wished that I could have had the chance to see a fetal elephant in the womb, as they are remarkably well formed from a very early gestation stage. Having witnessed a cull on a previous occasion, I saw many of these tiny, fully formed elephants being weighed in the aftermath. I wanted to see one that was happily nestled in the womb, with its cute little trunk, floppy ears, and upturned mouth, but fortunately for the experiment, the cows had apparently simulated pregnancy successfully. The method was then successfully used on twenty-three elephants on a small reserve outside Kruger, but for larger populations or for areas that are already considered to be over-populated, the method may be impractical.

A more recent experiment in 2005 posits that the most dominant bulls win the most matings. Researchers have performed vasectomies on a few dominant bulls. Vasectomies in a species with internal testes is quite a feat. Researchers expect that since most of the matings would have been performed by the now sterile dominant bulls, the birth rate will decrease.

In addition to these population-controlling experiments, there have been proposals by the South African government to resettle some areas of the park with desperately poor South Africans. There is talk of a transboundary park, taking down fences between the Kruger border and a park in Mozambique, but such deliberations come and go, and the conflicts continue in the cornfields and in reserves where managers don't feel it's appropriate to manage parks for the benefit of a single species.

I WAS ABLE TO GET high-resolution satellite images of the Caprivi region through the then Ministry of Agriculture, Water, and Rural Development, and a botanist colleague in Etosha had inspected each of the zones defined by contrasts in the images to confirm different forest types in each area, where a mixed dense acacia forest was darker than an open terminalia grassland, for example. Armed with this information, I selected twenty transect sites along the Kwando River, some near the river's edge and some 3 miles inland. Then I chose a representative mix of tree types and forest densities.

Tim's regular research schedule had been postponed because the satellite collars for his project didn't work and needed to be sent back to the United States to be refurbished. He would have to wait for the following dry season to get the collars on the elephants. So he and I had planned a quick safari to set up my transects and to do his dung counts.

Dung counts had been developed as a census method where the vegetation was too dense to enable aerial surveys. By clearing off certain sections of the roads and counting the fresh elephant dung, we were hoping to get a rough estimate of the population in the area. We were doing this during the first year, before there was money to do aerial censuses, but we later determined that the method was hopelessly inaccurate for our environment. I was also collecting dung to look at how elephant nutrition fluctuated seasonally.

We woke up that October morning and packed up camping gear, shovels, and food to head out to the field for two days. We were doing the elephant dung counts in the Triangle region and headed out to the Golden Highway, which bisected the West Caprivi in a seemingly endless stretch of white calcrete, corrugated in the dry season and treacherously slippery in the wet season. We were cruising along the bumpy calcrete when a huge herd of buffalo stood muddy next to the road. The herd quickly took off into the bush as soon as the pickup stopped next to it. We both started to count, settling on an estimate of four to five hundred and assumed that they had come from Guesha pan.

In a year with good rainfall, Guesha remained full for several months into the dry season. Tim pulled away as the herd disappeared into the dense Baikea woodland. We headed down the Guesha track to set up a transect that I had already selected in the open woodland. The Caprivi had huge tracts of Baikea, also known as Rhodesian teak, prized for its very dense red hardwood. I liked to start with the Baikea transects because elephants did not seem to like this tree, so they were the easiest to do. Getting a few transects done quickly helped make us feel better about how many remained. Some transects could take several hours, depending on the amount of elephant-favored plants they contained. These areas also tended to have more buffalo and thus more tsetse flies, making working conditions tense for many different reasons.

We stopped at another area, a mixed dense forest according to my satellite image. I needed transects of each forest type, first near the river and a second of each type located about 3 miles away from the river. The mixed dense river transects were the most utilized, since they were along the path to the water and they contained the elephants' favorite foods. The whole area had been demolished, particularly the *Acacia nigrescens,* which must taste like candy to the elephants, as the adult trees were almost completely debarked, the saplings chewed off like cane stalks and the seedlings eaten down to the base. No wonder this tree evolved thick thorns on its trunk (its nickname is "knob thorn"), and yet even the thorns were no match for the elephant.

We looked for a representative transect area, took a random compass bearing, and set up the transect line. It was important that our selection was unbiased, even though the devastation at this location was perfectly uniform. This area was particularly dangerous, however, as it was next to a lagoon, putting it in line with the traffic of many species going to drink, including buffalo and elephant. Tim kept the rifle on his shoulder just in case.

I looked at the fine thorns of the knob thorn seedlings and compared them with the adult form, a thickly barked and gnarled nipple protruding from the trunk, with a sharp claw sticking out, hooking downward. These knobs studded the entire tree trunk (in the few areas where bark still remained). It was fascinating to consider the evolutionary pressures driving the race between host plants and their herbivores.

Elephants have evolved relationships with many plant species. For some species of plants, the elephant gut is an essential first step for seed germination. In fact, elephants could be called the "Johnny Appleseeds" of Africa, responsible for the dispersal and planting of many of the continent's acacias. In the case of the erioloba, the guts of elephants and other large mammals not only soften the hard seeds and hulls, but also kill the beetle larvae that infest them.

When we finished our work for the day, we continued on to Horseshoe. At dusk, we rounded the corner to a view over a great oxbow lake where the river bent around in a perfect horseshoe. Another large herd of buffalo was just finishing their drink. A baboon troop quickly descended from its sausage-tree roost. Rather than climbing higher when threatened, baboons descended from a tree.

We set up camp among the hippo choruses, near the baboon tree, a large ebony that towered over the corner of the oxbow. Several hippos approached us from the water and blew air at us from their nostrils, flashing their huge teeth from gaping mouths as a warning to stay clear. We rinsed off with a quick sponge bath, anxious not to remain at the water's edge for very long. In this fairly deep water, a croc could easily sneak up to the surface. Knowing that the laundress from Lizauli had been eaten not too far down the river, we were particularly wary.

We heated up some cans of mutton and settled into the tent to read. I drifted off to sleep, the noise of the baboons right above us as they resettled into their roost for the night. An occasional squawk, probably from a troublemaking adolescent, would set off volleys of screeching and howling from the younger ones before an older male would have to step in with a deep, throaty reply, threatening them into silence. Tim and I were used to these nightly melodramas, since we had a troop that roosted just behind our house. Some nights we would awaken to the most horrific bloodcurdling screams, deep, thunderous roaring, and "hoo-haas" as the troop tried to fend off the resident leopard, on whose dinner menu baboons are a preferred item.

Sometime well after midnight, the oxbow lake filled with prehistoric wailing, roaring, and trumpeting echoing in the dark as elephant family groups reunited after three days of foraging separately inland from the river. With so many elephants in the area, we knew that it would be a difficult day of transects.

The following day, after several nerve-wracking transects, we left the Horseshoe area late in the afternoon. It was later than we would have liked, considering that it would take a least an hour to get home, and the dense forest at Doppies was not safe after dark if there were elephants around. We drove slowly down the track of Horseshoe, rounding the bend, which opened into a section of terminalia savannah where the trees were so heavily browsed that they looked like a grove of bonsai shrubs with thick trunks. I chose a waypoint for a perfect transect that we'd have to return to measure later.

After passing through the open savannah and entering the dense forest at Doppies, we were suddenly surrounded by elephants. Without thinking, Tim floored it, teeth clenched and eyes like saucers as elephants trumpeted and approached from all sides. It was like a scene from Jurassic Park. I began shouting immediately, "Slow down! Slow down!" I held up my arms too late and my head hit the roof as the truck bounced over a fallen tree that the elephants had dragged onto the road. More mad trumpeting burst through the bush as the matriarch closed in on the pickup. "Go left!" I screamed as the white of a tusk was about to strike the door. She swung her tusk at me just as Tim swerved the truck in the other direction.

"We're surrounded. Damn it!" Tim swung the wheel wildly as we veered back and forth between the trees for half a kilometer before the roaring and trumpeting subsided. The elephants slowly disappeared behind us as the truck slipped out of the forested area and into another open grassland.

"That was a close one!" Tim wiped his brow in relief and glanced over at me. I was fuming.

I had frequently seen how the rangers approached elephants by vehicle, accelerating toward the herd, aggressively speeding through, while the elephants charged defensively, trumpeting, roaring. The rangers then aggravated the situation by emptying their R-1 semiautomatic rifles into the air, scattering the herd into the bush.

"If you just drive slowly, they won't do that! You almost got us killed!" I was furious.

Tim was incredulous. "You've got to be kidding me! Don't you remember what just happened to those German tourists?"

Friends had just told us of helping these tourists, whose Land Rover was turned over and repeatedly tusked by an angry adult cow for no apparent reason. Fortunately, they had remained in the vehicle until she had spent her fury and left. Not realizing the dangers of hippos and crocodiles, however, they unwisely left the area by walking up the river and were rescued by the manager of a local lodge.

"Maybe they also hurtled themselves at the elephants, thinking that was the best strategy."

Tim now got angry at me. "So, what, now you're an expert in elephant behavior? I'm sure the rangers, if anyone, would know the best way to deal with them."

"Yes, but many of them didn't grow up in the bush. They were just stationed here recently, as adults. It's all-out chaos when the rangers drive through, so it's no wonder the elephants freak out. Look at how the elephants respond to shots fired from the truck. There's got to be a better way. If we gave the elephants the right of way and allowed them to pass peacefully, they wouldn't feel threatened by us."

I begged him to consider my alternative approach, recognizing that we were visitors in their space and should yield to their right of way. That night, however, we only got as far as a truce and headed on, knowing that there was one last patch of dense forest to get through before we got to the house. It was already dark.

Tim started to stiffen up as we crossed the Golden Highway onto the Susuwe track. "Great. You know the forest around Susuwe is bound to be crawling with elephants." They did seem to come in waves. Every few days, a new bond group paid a visit to the river. They would head inland to forage, then circle back to the river to drink; and there were large herds all along the Kwando. This was their movement pattern in the dry season, and it took about three days to complete.

"The only way to get through this section is to gun it. You know that. The rangers always do it."

I shook my head and offered to drive. This got Tim's blood going, but he did promise to try my technique in the future, in the daylight, experimenting in the floodplain rather than in the forest. There was no way I was going to win this one. But eventually, after several peaceful encounters with elephants, with me in the driver's seat, he reluctantly admitted that I could be right. After several of the rangers saw how I was handling the situation, they laughed and agreed that my method might be better.

We made it home without incident. There was not an elephant on the road until we pulled up the path to our house, which was dwarfed by elephants feeding on the *Acacia erioloba* pods scattered over the sandy clearing in front of the house. There were about thirty of them, and they made a run for it when our headlights broadsided them. Nonetheless, we waited a while before getting out of the truck and heading inside. Tim got out the maglight and scanned the clearing to make sure the resident leopard was not lurking somewhere. We grabbed a few essentials and headed into the house to shower, heat up some cans of beans for dinner, and go to bed, exhausted. The most difficult vegetation transects were now behind us.

AS NOVEMBER CREPT IN—the hottest, most humid part of the year—we finished the transects and switched into fencing mode. Not the best time to plan a community project, but the only time before the rains come, when everyone would be busy plowing and planting their crops. We had been camping with the game guards just outside Lianshulu village in a mopane scrub forest with no shade, mopane flies buzzing mercilessly at and in every orifice.

While on fence-building duty, we occasionally encountered the Mudumu herd, a large extended bond group of about two hundred fifty elephants. Elephants are usually slow and deliberate in their approach, but some are more active and aggressive than others. The Mudumu herd was renowned for its aggression, most likely because it is a small herd in a very small park surrounded by angry farmers.

They chased us unprovoked a few times, fortunately on the open grassy straightaways of the terminalia savannahs and not in the dense mopane forest of Mudumu National Park. When I looked in the rear-view mirror and saw ears pinned back, heads bobbing, and violent trumpeting, I knew the situation was serious. Sometimes, when flying through the tall grass in the truck, the elephants failed to appear any smaller in the mirror. Then my heart would start racing, especially if the path was blocked by the elephant's "Do Not Enter" sign, a large tree trunk dragged onto the dirt trail.

Once, from a distance, we had watched the Mudumu herd feeding. At first we could only tell where they were from the sounds of trees breaking, so we waited as they slowly browsed their way toward the perimeter of a tree island where we could see them.

An adolescent bull snapped in half an almost full-grown *Terminalia sericea* while a young cow dragged a debarked terminalia branch along with her as they headed for the dense mopane forest. When we could no longer see the herd, we reluctantly headed back to our uncomfortable camp.

Midway through the fence building, Tim and I went home to pick up more supplies, take a shower, cool off, and have a night to relax. There was no solace from the sun. The "cool" shower was almost boiling, the water in the metal tanks having cooked all day in the merciless sun. After scrubbing off the mopane clay, we sat outside and stared at the wavering floodplain as cicadas clicked their ear-piercing maracas.

Smoky wafts of hot air emanated from the river as the hippos squabbled over whose head would rest on whose back as their wallow shrank. Suddenly a swarm of locusts emerged in the mirage of a horizon, millions of wings providing the only cool air we had felt since the start of summer. We ran out on the open floodplain and stood in the middle of the swarm to marvel at it. We were engulfed in papery, beating wings.

A ranger came by later to tell Tim that he had come across a dead elephant just south of the station. There had been an outbreak of anthrax in the elephant herds that year, and this was the suspected cause of death. Tim had been monitoring the outbreak, taking samples from any carcasses we came upon on our patrols. Since we had been away for a couple of days, the ranger wanted to make sure that Tim got a sample of it before lions or even some of the villagers got hold of it.

The villagers were monitoring the outbreak by watching for the vultures circling overhead, waiting with knives and axes sharpened. They sneaked into the reserve and hacked away at bloated carcasses, bacteria exploding into the air as the gas-filled innards were punctured. Anthrax, as found naturally in the wild, does not compete well with putrefaction bacteria, so after killing an animal, its real success lies in its ability to sporulate quickly and remain in the soil for long periods of time. If a carcass is not opened, there is a better chance that anthrax will die before it can form spores.

We had come upon this particular young bull several days earlier, lying dead on its side next to the river, blood gushing from both the trunk and anus, fluid-filled vesicles on its stomach, all telltale signs of anthrax. Tim had already taken samples for the record.

We decided to go down and have a look at the carcass again, and sure enough, a crowd had amassed. We tried to warn them of the dangers of handling an anthrax-infested carcass, but they just laughed and said that they knew the disease and that it was not a problem for them. Seven tons of free protein was just too hard to pass up.

We weren't sure how much anthrax was in the air, but we didn't want to get too close to these enormous, bloated intestines and putrid flies buzzing back and forth between carcass and people. It was tempting to ask one of the men to slice up a foot or parts of the head so that I could have a look for my own research, but I thought better of it. Instead, I stood back and wondered about this poor elephant's final days. He looked old enough to have left his family group, but had he left any friends behind? As far as I could tell from watching them, the loss of a loved one seems just as hard on elephants as on people. In fact, not only relatives visit the graves of dead family members, but completely unrelated elephants have been documented visiting the graves of others. It's no wonder why there is so much lore around the concept of an "elephant grave yard."

The men hacked at the bloated remains with such fury that it was remarkable that they didn't slice an arm or a hand accidentally. Huge strips of meat were hung on long branches and carried across the men's shoulders back to the village. Probably because of how they prepared the meat by boiling it for hours and possibly aided by natural resistance to anthrax, the people weren't getting sick from eating the diseased flesh.

On our way back, we stopped in at the ranger station to report the event just as a local foreign aid worker was leaving. He had brought a very sick baby elephant to the station and didn't know what to do with it. Unfortunately, this happened often during anthrax outbreaks, as young mothers succumbed to the disease, leaving their babies orphaned. It was very difficult to get another family group to adopt a baby, because nursing mothers were not usually willing to share their babies' sustenance.

We crouched over the wheezing baby. It couldn't have been more than two months old. It could barely breathe, its lungs filled with fluid. We sat stroking it and tried to comfort it, but too weak even to lift its head, it died within an hour.

Another Caprivi orphan had been brought to Etosha a few months before, but it was in much better condition. He was about five months old, traumatized, missing the end of his tail, probably from a hyena, and didn't want to be left alone. He was just gaining control of his trunk, and he searched our pockets for food, pulling us toward him when he sensed that we were about to leave.

Since none of us had had any experience raising a baby elephant, one of the vets did a little research and learned what we needed to do to keep it alive. The baby required 4 quarts of a special warmed formula every few hours. I stayed with him once, and it was tough to keep on schedule through the night. He desperately wanted me to sleep with him, so I lay down with him in his hay bed until I thought he had fallen asleep. I gently got out from under his grasp, but by the time I approached the door of his stable, he had gotten up and beaten me to it and wouldn't let me out. He wailed and howled at my act of betrayal. If it hadn't been so frigidly cold, I would have spent the night with him, but with a final shiver, I finally worked up the strength to beat him to the door after three hours and many attempts to send him off to sleep by stroking his trunk and lying next to him. I ended up sleeping in the back of my truck next to his stable. I felt like a terrible mother.

Tragically, all these efforts were in vain, as he died a month later of diarrhea and probably a broken heart. I'll never forget his inquisitive trunk digging into my pockets. The poor little baby that now sat in my lap, dead, would not have had nearly such control. Another anthrax orphan down and my heart broke all over again.

Tim and I showered particularly carefully when we got home, since the odor of dead elephant seemed to pervade everything. And then it was time to get back to the fence. The clouds had started to amass once again. It was almost the start of another wet season.

7

THE MOTHER OF
ALL ELEPHANTS

*If you go through the high grass
where the elephant has already gone,
you don't get soaked with the dew.*

—GHANIAN PROVERB

THEY CALLED ME THE "Mother of All Elephants." The women
of Lianshuli village thought this was fitting since I had no children of my own, and I protected the elephants' food and
defended their land. These local women loved to tease me,
sometimes out of malice, sometimes as a compliment, and
sometimes out of pure wonder at the world of the white man
and his magic. When it came to elephants, however, the women
had only anger. Why would I care about how much nour-
ishment the elephants were getting from these women's mealie
fields unless these animals were my own? Naturally, all elephants must have been my children; it was the only logical
explanation.

The conversation with the women in which I got my name
started after one of them saw me coming from the forest carrying a hammer. She had been watching me nail electric fencing
insulators into a mopane tree in the forest next to her field. Hid-

ing behind a tree, she laughed hysterically with one hand over her sparse, tobacco-stained teeth. The other hand held a basket of false apricots on top of her head, one of the prized fruits of the wet-season forest. The women sitting in her field explained that a hammer was a man's tool. I thought this was ironic, since many of the men I had seen were often sitting under a tree stirring the pot of local politics or getting drunk at the Khuka shops, while the women were left to tend the crops and feed the children. Having once been a hunter-gatherer society, the agrarian lifestyle was a difficult adjustment for the men.

The women invited me for a break under the shade of their thatch and they opened a white melon to share. Then they divulged their nickname for me as they laughed at my curious interest in elephants. They asked about my work, interpreting through one of the game guards, and giggled at the thought of my measuring trees after the elephants had eaten them. They went into hysterics when I told them that I was collecting elephant dung to analyze it for nutritional content. I could see it in their eyes: Was all this really worth putting off bearing children? And who would pay for such a thing?

The role of women in this society quickly became obvious, and I realized early on that having a male interpreter was not a successful strategy when talking with the farmers. The women tended the fields day in and day out; the women, in fact, did most of the work in the society. The women controlled the family, the food, and what went into the pot. If I had wanted to interview an elephant matriarch, surely I wouldn't insult her by taking an adolescent bull along to do my interpreting.

After a particularly heated meeting where Margie accompanied me on a day of farm assessments in the village of Choyi, we both realized that we needed a different strategy. An old man kept standing up and yelling his complaints about things that happened in the past. I thought that if I could just speak to the women on my own, and through another woman, I would cover a lot more ground.

In November 1993, I hired a female interpreter, Janet Matota. The idea was that the interpreter would be an IRDNC employee and eventually assume a community leadership role for women, facilitating the management of natural resources principally used by women, such as thatching grass and palms used for basket weaving. There were many applications for the position scrawled on stained and yellowing notepaper torn out of a schoolbook and carefully delivered by foot or bicycle to Lianshulu Lodge. The brief essays were filled with assurances like, "I am health, happy and strong," or, "I am very much interested in animals and I am having good health." In retrospect, it was kind of ominous to think about a preoccupation with health: a foreshadowing of the coming plague of AIDS. After I looked through the applications, I selected Beauty, Constance, Loveness, Sioma, and Janet to interview, asking them admittedly difficult questions such as, "What do you think is the most pressing environmental issue facing the Caprivi today? What do you think can be done to help reduce conflicts with wildlife? How do you think life could be improved for people in your village, and what changes would you like to see in the area in terms of development? Do you feel conservation is important and why?"

Janet Matota answered the questions with a clarity that seemed out of context to her rural village background. She was a natural leader, and my life in the Caprivi changed dramatically after hiring her. We shared a sense of camaraderie that simply wasn't there with the men with whom I was working, and her work ethic was impeccable. I was excited about the potential of the position, and Janet seemed like she would provide a great window into her seemingly impenetrable culture.

It was easier for me to be inspired to help the women. They seemed more interested in helping themselves than the men did. Women did not have much power in the village, but they did exert a lot of pressure within the household, which indirectly affected decision making within the community. One of the rangers pointed out that it was good that I was working with the women to manage resources, since they were the ones who cooked and put food on the table and ultimately pressured their husbands to bring meat home for their pots. Because of this pressure, he felt that the women played a large role in the amount of poaching that took place in the region simply to feed their families.

The women had low stature within the community. They were expected to be silent and were not allowed to be khuta members or to attend a khuta meeting. When Janet spoke up at our first meeting to introduce the community resource monitor (CRM) program at the Sifu khuta, it opened eyes and defined her as a leader from the very beginning.

At the time, there was a great deal of squabbling between the Mwfe and the newly seceded tribe, the Mayeye, which recently appointed their own chief and hence had their own khuta. Historically, the Mayeye had been the slaves of the Mfwe, so having a separate chief was particularly symbolic. There was a constant power struggle between the two khutas, most recently over the new khuta's wanting the community game guards in their jurisdiction to be placed under their authority and not that of their old chief, Chief Mamili. They viewed this action as an important recognition of their independence from the Mfwe. The leaders were preoccupied by this seemingly petty desire, which overshadowed the unveiling of the CRM program to this khuta. As usual, local politics had their way of crippling momentum.

Garth and Margie, the heads of IRDNC, and Matthew, their field officer, arranged for Janet and me to join their meeting with the Sifu khuta. Margie thought it would be a good time to introduce the CRM program to the local headmen, or indunas, within this forum.

We arrived at the khuta, where seven headmen sat outside in the shade of the open-walled building on tattered vinyl chairs, refugees from a seventies kitchen. Several held canes. As worn as their clothing looked, it was their best attire. They presented us with chairs, and several younger men sat under the trees on stools.

The meeting had been called by the khuta to discuss their desire to be acknowledged and respected as a separate entity with power and rights equal to the other khutas. They knew that the directors of IRDNC were not often in the area, so they took the opportunity and requested a meeting. Unfortunately, this agenda eclipsed any other issues, and for me it was just another reminder of the multitude of challenges facing conservation in the region.

In the absence of the chief and the second in command, the ingambella, one of the khuta indunas began the meeting with a list of grievances. "We would like Matthew to report to this khuta," he said. "The president himself had come up to the Caprivi to recognize this khuta, yet Matthew has chosen to ignore us to work with the enemy." The induna shifted in his chair. "We want you to surrender all the names of the game guards in this region to the Sifu khuta. We are now in charge of this area. We want that list under our command."

The induna sat back and gestured for the floor to be given to Matthew. Matthew sat up and cleared his throat. He lit a cigarette and took a long drag. "I have been working in this area for five years." Matthew looked at the khuta members. "Even now, I find it difficult to reach all people. There are problems working with all people because of tribal issues." He flicked his ash. "And I apologize for any mistakes, but I need guidance on this issue of tribalism, not just from IRDNC but from the communities." He took another small drag.

"Chief Mamili went to Johannesburg in 1988 to discuss the issue of national parks. While he was there, he learned about the game guard program that Garth and Margie started in the Kunene region and requested that that program start in this region." Matthew cleared his throat. "In 1990, IRDNC came to meet with the Linyanti khuta to discuss the initiation of the game guard program." He paused, gathering his thoughts. "At that time, some communities were not interested in the program. IRDNC worked with the only five communities that requested game guards."

He took one last drag, crushed the butt on the ground, and put it in the breast pocket of his battery acid–eaten khaki shirt. "We would like to explain the new post that IRDNC has created. Janet Matota has been hired as a community resource monitor. Until now, we have only addressed the men in the community. Women are using resources from the forest and do a lot of work in the fields. We would like to bring local women into our program and get their input." Matthew paused and looked down the line of khuta members for a reaction. "We are learning about wildlife in the area, but we know little about the people and how they live. The CRMs were not appointed to police the women on how they use resources but for us to get more information and ideas of how people use and manage natural resources."

When the forum was handed back to the khuta, it was clear after a drawn-out rebuttal that the only thing that mattered, aside from the complaints about the lack of development, was the surrendering of the game guard names to their khuta as a recognition of their independence and separate jurisdiction. The floor was then turned back to us, and Garth took up the platform by addressing the issue of development, a less contentious point. His pipe was finished and he knocked out the ash, loaded it, and lit up again.

"Margie and I are the directors of this organization." He drew from his pipe and exhaled. "We are an NGO [nongovernmental organization]. Margie and I are both Namibian. We don't have sufficient funds to pay for game guards ourselves. We have to look for these funds from groups outside Namibia, like the WWF [World Wildlife Fund], which wants to see wildlife protected but also knows that focus should not just be on the protection of wild animals. People living with wildlife should also live well." Garth squinted at the khuta members, taking another drag through his teeth. "They know that wild animals cause problems, and it is important to look at the problems so that solutions can be found." He paused again. "They also know that wildlife is not plentiful. Many people want to see wildlife. They want to pay to see it. That brings money to the country. The WWF sees money going to the government and to tour operators. They believe it is fair that people who live with animals and protect them should also get money and they are working with the government to address this issue.

"Development is important. IRDNC must work with everyone, because development is not just for some people; it is for everyone, men, women, and children. That is also why we would like to work with women. That is also why we have an environmental education officer to work with children. Tomorrow they will be the adults. We are planning an environmental education center at Sachona so that teachers and children can learn about wild animals. Then they can go and see the wild animals. Many children have not seen wild animals like we have when we were young."

The head induna stated flatly, "We still have a problem with the names."

Janet spoke up, frustrated by the childish behavior of her khuta. "You know that there will be a problem with taking those names because it was Chief Mamili who signed for this company. He will not want to give the names."

The khuta members were instantly furious, hissing and clucking their tongues at her. How dare she interfere with khuta affairs?

"No! You are wrong to say that because now there is a new chief," the induna scolded her. But, Janet had robbed them of their cocksureness, and they felt it bitterly.

"We want to work with all people, but we must be careful," Matthew repeated. "If changes disrupt the program, funding will be questioned."

"Let me give you an example." Garth banged out his pipe and refilled it as he continued. "Let's say we decide to build a new house. Young people come in and put sticks in the ground and quickly build a house. But if that house is built too quickly, it will fall down. We know as old people that in order to build a house, we must plan, and the sticks must be strong and go in the right places."

The khuta members did not want to be reasoned with. Garth persisted, holding his hands out. "I am asking the indunas to let us build this house carefully so that it is strong. If we build too quickly, it will fall down." He lit his pipe. "I understand what you say, but please be patient."

"I am trying to ask about the names," the Induna tried to twist the issue. "I didn't know we still fell under Mamili. This is the area for the Sifu." Names are extremely important, representing jurisdictional power, authority, and autonomy.

Garth tried another approach. "Let's say there are two people fighting, maybe man and wife. When fighting, blood is hot and the people are very cross. Old people must solve the problem. If you try to solve the problem when the blood is still hot, you will not solve that problem on that day. Better to wait and separate the two and see when the fighting stops. Then you will solve the problem."

The induna fumed. "As you know, long back, SWAPO [South West Africa People's Organization] party was fighting with SADF [South African Defense Force], but now it has been reconciled. Chief Mamili is the enemy. One of our people has been killed and one injured. I do not understand why you operate there because they are the enemies. Maybe it is okay to work with the Subia because they are our friends."

"I see your blood is hot." Garth looked at the induna through his pipe smoke. The induna did not respond. "If there is fighting, funding may stop. It is very important that we do not fight. Then we can move forward."

The induna tried a new angle. "Sorry. I just want to understand something. Are Bukalo names here or at the Bukalo khuta?" Some of the other indunas smiled. "We ask for names. It seems you are afraid about the trouble. You must not worry about trouble. We ask you to provide the names, but they can also remain with the Mfwe khuta." Some khuta members were not happy with this. "So, any problems here must be reported here to Chief Sifu as well. We cannot force you to take the names back now. After all, it is the Mfwe who fights, not the Mayeye." And with that, the induna made a few closing statements and the meeting ended in a thankful prayer.

After dropping off Janet, Matthew and I headed back to Susuwe. It had been a very long day, and the khuta meeting had left me feeling empty and cold. Local politics usually shifted the focus away from conservation. Angry clouds threatened an early end to the day, which came as an undeniable relief. We decided not to make a stop in Choyi to check on our fence, and instead Matthew dropped me off at the barracks just as the storm approached.

"Hell, I hope I can beat the rain."

"Looks like it's going to be a big one!"

We said our good-byes, and just as I stepped inside, the sky broke open and water poured from the heavens. I watched the rain and tried to remain positive despite the day's events. Janet's courage and tenacity gave me the strength to stay hopeful.

OVER TIME, the women's program grew, and, in partnership with IRDNC, I hired a colleague for Janet named Loveness Shi-ita in late 1994. We spent our time monitoring elephant damage with the women farmers, maintaining the electric fence around Lianshulu, and helping the villagers to improve and manage their resources so that they could sell better-quality products and farm more sustainably. The ultimate goal was to get people to recognize the importance of wild places and to adopt ways of managing them so that both people and wildlife could prosper.

One of the most successful of these sustainable products was the thatching grass sold to commercial vendors in Windhoek. The manager of Lianshulu Lodge had initiated the project. He was eager for native people to see the benefits of taking care of the wildlife in their region.

In the early days, I helped him set up the business. We held meetings to help organize the communities, get input, and explain the process and some of its limitations. Each woman was allowed to sell only two hundred bundles so that as many women from as many communities as possible could get money when the truck arrived from Windhoek. There would only be a few trucks per season, so it was important to be sure that the opportunities were evenly dispersed, allowing as many communities as possible to receive income from this venture. This plan, of course, required the cooperation of the thatching companies; it would have been much easier to load the truck at the first couple of grass stands rather than have to stop at each concession and buy two hundred bundles from each woman.

Our venture was not without a few problems in the beginning, a missing sickle here, a missing cash box there, arguments between villages over where the truck should stop next, and so on. Janet, Loveness, and I were there to settle squabbles and make sure that all the money went to the right places.

After one of the meetings that the lodge manager joined, he complained that he looked like the Big Bad Wolf and I like Little Red Riding Hood coming in to get all the credit, as all the villagers' anger was directed at him. Grant smiled as he said this, however, looking every bit like a rough-and-tumble Nick Nolte. He had a single-minded determination to help the communities, but in his eagerness to get things right and see ventures succeed, he tended to do things his way. He had set up a very successful traditional village so that tourists could see how traditional life was led in the Caprivi. But in the end, the community saw him more as a boss than a partner, and everyone blames the boss for his troubles.

We once went out to inspect the quality of the grass bundles in the community next to the lodge, and Grant became incensed with their poor quality; some were too small, some not packed densely enough, and others made with inferior grass. He went through each of the women's racks, pulled out the unacceptable bundles, and threw them in a rage, frightening the women.

I suggested that he come by with a sample of what he thought the vendor expected to purchase. I gently replaced the bundles and smiled at the women. Grant suddenly stopped and pointed at one of the women's racks, which contained the perfect product, all the bundles tightly packed, the right size, bound with the right twine, and consisting of the right species of grass. He was able to get his point across. The other women smiled and nodded in agreement that this woman had the most beautiful bundles, and they were eager to follow suit.

After the next grass pickup finished, it was dusk when several old women grabbed my hand, held it open, and spit into my palm through blubbering lips. It was a strange scene, with calcrete dust in the air mixing with the little smoky fires next to the road marking the racks of grass to draw the vendor's attention. I smiled nervously and asked Janet and Loveness what the women were doing. Loveness laughed and explained that the old women were showing their appreciation, thanking me for my kindness and for helping them. They were simulating the sizzle of a fire when water is thrown on it, the sizzle representing their ancestors, who were smiling on them through my good deeds.

I drove home with tears in my eyes, elated that such a simple gesture could have instilled so much hope in these women. When I got home, Tim was sitting in the middle of the floor surrounded by nuts, bolts, and belting material. The elephant collaring was about to begin.

8

ELEPHANTS
ON THE MOVE

Elephants carry "wisdom sticks" on either side of their temples. These sticks enable them to know the time and place of their own death. That is why very old tuskers are often seen without their herd, preferring to find a hiding place to die, thus maintaining their dignity, as they wish to die alone and in peace.

—SHONA LEGEND

THE THUDDING AND CLACKING of chopper blades cut through the air around us. After the collars used for elephant tracking had been repaired and tested, they were finally ready to be put on the elephants. Tim had all the collar components and tools organized and had planned everything with military precision. We started the operation on the Nova floodplain to the east, then moved to the Golden Triangle, the Kwando floodplain, Horseshoe, and, finally, Mudumu and Mamili National Parks. It took about a week to collar the matriarchs from ten different herds.

Each was fitted with a satellite collar, which consisted of a thick 4-inch-wide leather belting material with a satellite transmitter mounted on top, weighted at the bottom to keep the transmitter in place. There was also a small radio collar looped through the bottom of the heavy satellite collar belting so that we could find the elephants by vehicle as well as by radio tracking. In total, the collar weighed about 25 pounds.

As they were being collared, herd members demonstrated startling and sometimes heartbreaking bravery. When we flew directly over a herd to position the chopper over the matriarch so that the vet could dart her, some of the herd members would stop, turn around, and try to swat the giant dragonfly out of the sky with their trunks. Then there was the noise of the dart being fired, and the chase continued.

After a few minutes, the darted elephant stood out as the one being escorted by the others. As the drug took effect, the elephant would falter, which the herd members were quick to notice, and they would almost lift her off the ground with their tusks to help her keep pace. The little ones scrambled amid the panic, trying not to get underfoot. When a darted elephant finally went down, the pilot often had to buzz a youngster to chase it away, most likely a baby refusing to leave its mother's side. Some of these calves acted with inordinate valor, head up, ears out, and mock charging the helicopter, refusing to leave the scene. But once they were pushed back to a safe distance, the chopper landed and the team immediately got to work fitting the collar, while the vet tended to the elephant to make sure that she was comfortable, sometimes shading her face with her ear or splashing water carried by the ground crew on her when they were able to get in to the site where she went down.

We became very concerned during a pursuit close to Horseshoe, which took so long that one of the cows regurgitated water from her pharyngeal pouch, where elephants store many gallons of water. They use this water in an emergency, throwing it over their backs to keep from overheating. We were so far off the road at that point that the ground crew could not keep up. But having already isolated the small family group away from a much larger extended bond group, it would have been more stressful for the entire group to retreat and break out a second family, so the vet got the elephant down as quickly as possible to avoid any further disturbance.

The pilot once had to leave us to refuel the chopper while we waited for a cow to wake up. She was breathing so heavily through her trunk that she was snoring. In a touching display of concern, one of the rangers put a small stick between the two lips of her trunk to prop open the airway to help her breathe easier while we waited.

It took so long for the chopper to return that the vet deemed it unsafe to leave the elephant down any longer and gave the antidote before the chopper got back. Then we were there with the elephant, in a wide-open floodplain with nowhere to hide. We were hugely relieved when the cow decided simply to stumble off in the opposite direction and not take us to task.

At the time of each collaring, we cut small notches into one of the tusks using a hacksaw. In two years, when the collars would be taken off the elephants, we could use these notches to measure how much the tusk had grown from the base of the sulkus (the skin surrounding the tusk). Ivory continues to grow throughout an elephant's life, and we wanted to compare the growth rate for the Caprivi elephants to the rate in other places. We also took a small piece off the tip of the tusk to save for future studies. Researchers had shown that information extracted from elephant tusks could facilitate determining the origin of the elephant. They hoped that this could be used to track illegal ivory or as a method for regulating legal ivory.

Once we had completed the collaring, we settled into a routine. Sundays became radio-tracking days, as the satellite collars were programmed to turn on and emit a signal on Sundays. That signal was detected by satellite, and the information was then transmitted to researchers in Etosha National Park. The radio collar transmitted a radio frequency signal that we could pick up using a radio antenna and receiver and thus locate the elephant. These sightings became a ground reference for the satellite data. Tim could use this information to know how much error to factor into the satellite data for elephants that we were not able to track or elephants that had traveled too far into the Okavango delta or up into Zambia. These early satellite collars had much more error in them than modern ones we use today. Now, we can even text message a location to an accuracy of within several yards. Tim was determined to get the best estimate possible with the available technology, though sometimes the error spanned miles. But since nothing had been documented about the movements of this multinational population of elephants, even a gross measure was important.

Often on Saturdays, we would stay overnight somewhere in the southern tip of the Golden Triangle and then proceed up through the region on Sundays, when the collars were transmitting. It was exhilarating work, although our flimsy nylon tent began to feel a little meager and uncomfortable when a herd of buffalo stampeded next to us on the way to the river in the dark. We also heard the leopard calls and lions prowling around us. After a while, we decided that a platform at Horseshoe would be a much more comfortable alternative. Given our meager budget, it was a cheaper solution than a proper canvas tent, and it would give us a spectacular view of the elephants as they poured into the oxbow lake at sunset.

Occasionally, in the laze of a Sunday afternoon, we combined our radio-tracking mission with a wallow in the shallow backwater just above Hippo Pools. Once we joined some of the other ministry staff in a communal effort to keep cool. There were a few shallow, clear backwaters that the rangers had deemed safe to lie in; that is, they were too shallow for a crocodile.

During our wallowing, a herd of fifty elephants approached hurriedly, stopping occasionally to smell the air, but we were downwind and they weren't paying attention to us. Not wanting to startle them, we remained frozen in the water as they skimmed the surface with their trunks. Unfortunately, we were also downstream, and in their excitement to reach water after three days of foraging inland, the elephants promptly urinated, defecated, and churned up the bottom while drinking, which didn't seem to make any sense because elephants normally prefer to drink clean water. It wasn't until the matriarch strolled toward us and almost stepped on one of the rangers that she became aware of our presence.

Both ranger and elephant were equally shocked as the elephant tiptoed briskly backward, trunk held high, ears out, eyes wild until she reached the herd and led them quickly but silently back into the forest. We sat in the warmed water until we felt it was safe to return to our truck. Elephants seem to prefer that humans remain in vehicles.

Another time, Tim and I combined some tracking with wet-season vegetation transects that were carefully selected, since only certain roads were navigable in this season. Our favorite route was along the Malombe pan road and then down to Guesha. It was difficult to see elephants in this season, as they spent most of their time foraging and grazing inland, drinking from water that collected in pans in clay areas of the forest. We occasionally saw them along this route, drawn there because of the sweetgrass near the Malombe pan, a favorite wet-season staple.

We sometimes got to see a bull guarding a cow with her small family group as they dined eagerly on the long grasses. This three- to six-day period is the only instance when adult bulls spend continuous time with a family group, wanting to be sure that the female is safe during ovulation. Since females only come into estrus once every four years and have a gestation of twenty-two months and a nursing period of two to three years, it is in a bull's best interest to stave off the competition to ensure that he sires the next generation.

As we drove farther down the track, the tsetses were amassing into a black cloud behind the truck. The tsetse fly numbers were reaching epidemic proportions in the West Caprivi since the SADF had left and stopped maintaining the traps that effectively controlled the population. They were worse on driving patrols than walking patrols because they tend to follow large, fast-moving objects, seeking out blood with the bite of hot needles. This became a serious hindrance to our patrolling until I made screens for the windows with mosquito gauze and Velcro.

Then when we arrived at our destination, we'd have to sit still and wait until the swarm died down. Once we got far enough away from the truck, the numbers were more manageable and we'd get only a few bites.

Although the human form of African sleeping sickness, caused by a parasitic organism carried by the tsetse, does not occur in this region, domestic stock and pets are vulnerable. My immune system was just starting to get used to the horrible little beasts, but in the beginning they made huge, painful welts on my arms and legs. Yet as vicious as it is, the tsetse is credited with keeping large tracts of Africa wild. Tsetses are the reason many national parks were established across Africa; once traditional hunting areas, they were never populated because of the threat of sleeping sickness to domestic stock.

AS THE RADIO-TRACKING MISSION was in full swing, money became available to do an aerial census of the region, involving total elephant counts along the Kavango and Kwando Rivers and additional counts inland within the West Caprivi Game Reserve and the Mudumu and Mamili National Parks. Tim was in charge of the census design, booking the airplane, and transporting the fuel to the local airstrips, as well as software and hardware integration and whatever statistics were needed. Tim loved this work, and Jo and I acted as backup.

During the Kavango counts, we stayed in cabins at Popa Falls, a government-run campsite just next to the Kavango River. Although not heated, the cabins made us happy, as it was bitterly cold at night, particularly next to the river. Tim and I set our alarm clock for 4:00 A.M. to take motion sickness tablets and then sleep for an hour and a half before having to wake up

"Young bulls are full of mischief and will often chase other animals away
from the waterhole, head up and ears held out to make themselves seem
as intimidating as possible, even when chasing a flock of guinea fowl."

"I had spent a few nights in the back of my truck initially, in order to scout out the site, but the lions were getting so bold that they climbed up on the bumper to smell me inside my tarp-covered pickup."

"I spent most of my time in the hide, a seven-foot square cement bunker 30 feet from the water, with about 6 feet of its height buried in the ground and a pillbox slit facing the waterhole."

"Kevin was just coming out of musth . . . with classic, exaggerated trunk curls over the head and then his signature dragging about two feet of his trunk on the ground. He would also swing it in Willie's direction, who in turn contracted his trunk so that it fell just above his knees, pushing back, as if hoping Kevin would stay put and stop approaching him."

"I stared out from the Mushara tower over the chilled landscape lit by a three-quarter moon. The unusual cloud cover obscured my view, smothering all but a few shafts of light that stretched out to some far off place in the east."

Planting a shaker, the device that delivers vibrations to the elephants. It is housed in a plastic dome to protect it from the sand.

"Back and forth the two-step went, a trunk push and two steps back, then another trunk shove with trunk overhead and back two steps in the other direction."

"Although mostly subtle, Greg must have been kicking a lot of butt behind the scenes to get some of the reactions that we had seen at the waterhole, where he calmly commanded his reign over the trough."

"[The baby] suckled awhile with trunk hanging to the side, just next to its mother's front knee. (A nursing elephant is an incongruous sight because the mother's teats are located in the chest area. It always reminds me of their close relative, the dugong, whose head and chest poking up through the long sea grass led to the tales of mermaids.)"

"The cycle of elephant movements gave the bush a marked rhythm that bound us to its music. We soaked it in, drunk with a love for the land that ran so deep and strong that it scared us. We had discovered a life that we couldn't bear the thought of leaving."

for the census at 5:30. The type of flying needed for a census in a small airplane required sharp banking in order to follow tight transects—brutal on the stomach. At times, I felt that my stomach was actually lifting and turning with the airplane, independent of my body.

Jo had all our dinner menus planned out for three days and brought along half a steer's worth of vacuum-packed meat to store in the camp freezer. During a particularly memorable dinner of slowly barbecued lemon rosemary lamb ribs, Tim told Jo what it took for us to get motion sickness tablets out of the Katima clinic. Jo reeled hysterically at how we had convinced the local doctor that we knew we were going to get sick, even though we weren't actually sick when we went to the clinic. Preventive medicine was not often considered in these parts.

During our transect counts, Jo kept my sickness at bay by feeding me slivers of wet biltong, the local style of dried meat that was still moist inside. I never thought that such a thing would be enticing, particularly in the early mornings, in the back seat of the Cessna-182, but surprisingly it came to my stomach's aid.

Up and down the transect lines we flew, carefully following the coordinates that Tim had set for the count. We marveled at the huge, extended elephant herds, contiguous groups that seemed to go on for miles as they followed the river. Occasionally, we had to dodge vultures circling in a warm air current.

At the top of the Kavango, we had to bank into Angola, and once we thought we saw a poachers' camp, racks of meat freshly smoked, smoldering cooking fires not more than a day old. Tim got the coordinates, though it was out of Namibian jurisdiction. We tensed every time we crossed the border, expecting the Angolan military to come over the radio and demand that we identify ourselves and explain our presence. We hoped for that minimum level of decency rather than shooting us out of the sky as the Botswana Defense Force threatened to do on a later flight when it came time to remove the elephant collars.

One night, Jo grilled pork chops, entertaining us with more stories about Katima. He told us about the last time that Kai went to get his truck serviced at the government garage. "Can one really stay sane in a place like this, I ask you? Get this. Kai takes his vehicle in for servicing on Monday. It's now Thursday. He goes back to check on it and he finds it on blocks. No tires."

Tim laughed.

"Yes, Kai drove the vehicle into the garage, and now it has no tires."

"What did he do?" I smiled encouragingly.

"Well, he walks up to the person at the front desk and asks what happened to his vehicle. They are waiting for more tires. 'But I drove it in here with tires!' Shrug. 'No tires.' 'But I did not come here for tires. I came for a minor service.' 'Sorry, I cannot help you. The tires are on backorder.' 'Backorder! Backorder!' Kai says, 'You bloody find my tires and put the bastard's tires that you stole from me on back order.' 'Sorry, waiting for the supply truck from J-berg. We don't have tires to fit your vehicle.' 'But these are standard 16 mm tires! All of these tires are the same size.' Blank stare. 'Sorry.' So Kai storms out of there with steam coming out of his ears. Can you believe it!"

"Jesus, Jo, I don't know how you take it!" Tim shook his head.

"I can't. I can't take it. I must get myself to Windhoek. God's country is full of broken dreams."

LATER WE CONDUCTED several censuses using a Cessna-337 used in Vietnam that was donated to Namibia by the U.S. military. We were delayed by student pilots flying planes to the Caprivi and forgetting to put the landing gear down before landing. Needless to say, we had to wait for yet another plane and clearing goats off the airstrip seemed a more preferable stumbling block.

As the seasons passed, wet and dry, wet and dry, we watched

the elephants pour in and out of the Caprivi in the opposite flow of the mighty rivers to which they were bound. They flooded in along the dry-season banks of the Kavango, Kwando, and Zambezi, away from the bleak withered woodlands, and as the rivers began to swell with the rains of Angola, the elephants made their trek back into the lush green forest.

The cycle of elephant movements gave the bush a marked rhythm that bound us to its music. The elephants were like silent conductors of a natural orchestra, seeming to summon the frenzied frog calls of the wet season and the raucous hippo bellowing, all of which reverberated up and down the river at sunset. We soaked it in, drunk with a love for the land that ran so deep and strong that it scared us. We had discovered a life that we couldn't bear the thought of leaving.

9

CRACKING ELEPHANT MORSE CODE

The torn boughs trailing o'er the tusks aslant,
The saplings reeling on the path he trod;
Declare his might: our lord the elephant,
Chief of the ways of God.

—RUDYARD KIPLING, *The Jungle Book*

AT THE END OF OUR THREE-YEAR CONTRACT, when it came time to leave the Caprivi, I was stricken yet freed. Which way did I feel? Which way should I go? How could I tease apart these feelings? I had exhausted myself with indecision. I had never experienced such a funk before, and yet I never felt more alive.

How is it that I had come to grieve for this land, for the animals, and for these people? How did I let it consume me? How could I put things in perspective? After leaving and gaining some distance, would I ever be able to return? I wanted desperately to help, yet my visions for the inevitability of failure paralyzed me. In the end, had I really helped these people? Had I made a difference? Or was I walking away with the rewards, leaving them to carry on, still in need, and perhaps painfully more aware of the First World's riches?

The night before we left our little thatch house at Susuwe Ranger Station for the last time, Tim had gone to the ministry office to say good-bye and found everyone in a flurry of activity. They were loading their weapons, preparing for an engagement. UNITA (União Nacionalpela Independência Total de Angola) forces were heading our way. Amid the chaos of international politics, somehow the Angolan government was given permission to operate against UNITA on the Namibian side of the border. This was the final blow, as UNITA soldiers moved their poaching camps closer and closer to Susuwe Ranger Station, gaining courage to challenge the station. Living 8 miles from the Angolan border had posed similar threats in the past, but this time it was serious. The Namibian army gave up patrolling the border for poachers, and UNITA was spreading its wings.

That night, we slept with a rifle under the bed, half awake, waiting for the inevitable popping of automatic weapons, running and shouting in the distance. The noises never came, and we left the next morning in a quiet fog. It was comforting to think that we left because we had to, not because we quit. Or at least I tried to believe that, knowing that the land and the people would always withstand the ravages of war. They had nowhere else to go. It was merely the soldiers and the foreigners who served as a changing of the guard.

Returning to the States after four years in Africa was a difficult adjustment—so far away from feeling the natural rhythms of the earth. But Tim and I were forced to adjust quickly, as our Ph.D. program at the University of California–Davis started in less than a week. I had to adjust on many levels, socially, psychologically, and physically. And I could tell that I could slip back into the First World and its comforts too easily, forgetting about the struggles of the people and the animals that I had come to know and love.

I found myself reassessing where I was going in my life and how I had gotten to where I was. You would think that I would have had plenty of time to ponder such things in the bush, but taking stock of myself was not a luxury I was afforded in a position that mainly involved crisis management.

I had thought in my youth that I wanted to become a doctor, but now I was starting a Ph.D. program in ecology and wanted nothing else. When it was thought that I might have a hearing problem at an early age, I had become fascinated with ears. Little did I know that when I followed around my doctor dad, carrying his medical bag, wanting him to look inside my ears, this early interest would form the foundation of my future career studying hearing and animal communication rather than clinical medicine.

It was my dad who introduced me to the wonders of the natural world. In a sacred way, he revealed the first crocus of spring, removing the debris around the jack-in-the-pulpits in the forest as he chopped a fallen birch for firewood. He reveled in the salamanders and crayfish in the brook and marveled over where the deer must have fawned in the fairy circle next to the old barn.

I ran my early wilderness operations out of a hut that he built for us in the woods. There I learned about the plants I could eat and from which I could make medicines. Having grown up on a farm that was founded more than three hundred years ago by Dutch immigrants to New Jersey, some of my imagination was inspired by an ancient Native American burial ground that existed prior to the Dutch settlement. My budding interest in the natural world also included a frog capture-and-release program that I set up at the stream next to a neighbor's house. Not surprisingly, I was the only willing participant.

I realized that a degree in biology could lead to a more appropriate career choice for me than being a doctor. After completing my bachelor's in science, I had the opportunity to volunteer on a lizard ecology study in the British Virgin Islands. There were actually people out there who did that fieldwork for a living. I wanted this life more than anything.

At the field station, I saw firsthand the passion that two entomologists had shown for their work. I followed one of them to Hawaii, where I worked at the Bishop Museum, learning about the wonders of island biogeography and the amazing ability of insects to evolve. I earned my master's in entomology doing what I had always done best, watching animals, completely absorbed with their behavior.

Soon, however, Africa cast a spell over me. I thought I had known what I wanted, but I became completely enraptured by Africa, so much that it scared Tim. Having grown up in South Africa, he was searching for something different, away from where he grew up. The Namib dunes were a dream for me, but for Tim, Namibia was where he was to be sent as a conscripted soldier had he not left to pursue a degree in the United States.

Back in an academic environment, I was determined to explore my theories that elephants might use seismic waves to communicate. I also wanted to explore whether seismic deterrents could mitigate conflicts between farmers and elephants. First, I needed to prove that elephant vocalizations produced seismic waves that spread through the ground. And if that was the case, did they spread over a distance that would not only add to the sound itself but perhaps facilitate further travel, which would increase the distance over which elephants could communicate? I would need to demonstrate that elephants could, in fact, detect these seismic cues.

I discussed the idea with my Ph.D. adviser, Lynette Hart, and learned that her brother, Byron, was a field geophysicist. We all had our own reasons to be fascinated with elephants, so we soon made plans to design a set of experiments. After a few preliminary trials in zoos and private facilities in California, it was clear that the noise in the ground at any of the available sites was going to be a problem for our studies.

We eventually found a private elephant reserve outside of Jefferson, Texas, that seemed perfect for our purpose. Byron had designed a traveling geophysics unit in the back of an old white van. His plan was to strike it rich using his seismology skills to find gold on his claim outside of Winnemucca, Nevada. He knew how to measure lightning strikes and use them as a type of sonar device to understand the composition of the rock in a particular area, reading the earth to find gold.

With our gold diviner in tow, Lynette and I set off for Jefferson. Thunderstorms were brewing overhead as we made the long drive from Austin. Byron entertained us with Gilbert & Sullivan favorites and tales from the field along the way.

Byron was just the kind of thinker we needed to help overcome the enormous technical challenges that our studies posed. When he recorded our first elephant rumble in the field, he got so excited over how elephants produced the vocalizations at such a low pitch that he simply blurted out, "If only elephants had lips on their ears, they could seal off everything aboveground and focus on everything below to detect the vibrations better!"

To our surprise, with a little probing of the fleshy area around the ear canal, we quickly realized that elephants did in fact have the ability to voluntarily close off their ears. Without realizing it, Byron had made a major discovery. We later learned that this unique anatomy had been described at the turn of the century, but a function had never been proposed. We characterized the muscles surrounding the ear canal opening as being similar to those of the eared seals, with skeletal muscle surrounding collapsible cartilage, forming a spiral that closes off the ear canal. For marine animals, this adaptation serves to keep water out of the ear; for the elephant, it may have served the same purpose while crossing a river, but we began to suspect that it may have an additional purpose if elephants were tapping into the ground as a mode of communication. Also, sealing the cavity might form a closed acoustic tube, where the pressure built up by closing off the canal could facilitate the movement of the middle ear bones, increasing sensitivity to the vibrations.

Other researchers had shown that elephants' ears were specialized for low-frequency hearing and that they could produce, detect, and respond to low-frequency rumble vocalizations. We needed to prove that while these very high-amplitude vocalizations in the range of 20 hertz were produced in the air, an independent wave containing the same information also traveled through the ground.

Unfortunately, the Asian elephants at the study site in Texas weren't particularly vocal, and it was initially hard to get any data. We had parked the van near the elephant enclosure, anticipating their rumbles, finger poised over the button to trigger the recording system. The system wasn't exactly perfect. Sensing our frustration, one of the keepers suggested a solution: he knew how to get his girls excited. Tossing a flustered chicken under the fence, we watched the raging charges ensue. From this, we discovered something else, which in retrospect isn't all that surprising: elephant charges create an enormous seismic wave.

In between the charges, gleeful roars, and trumpets while chasing the unsuspecting chicken out of the enclosure, the cows produced several low rumbles, which we were able to use to show that they also spread through the ground as a separate signal at a different speed. In theory, if elephants could detect these seismic signals, they could be detected at much farther distances than airborne sound, which could play an important role in expanding the range over which elephants communicate.

The waves in the soils of Jefferson, Texas, traveled much more slowly than sounds that traveled through the air. But because the speed of vibrations traveling in the ground depend on what the ground is made of, the first thing we had to do was establish the velocity of vibrations at our study site. I soon recognized a common instrument among field geophysicists: a sledgehammer. In order to measure the velocity of soil, we could just pound on the ground with a sledgehammer.

Of course, we also needed geophones and a triggering device, set to go off once the ground was struck. This simple action allowed us to measure the time it took for the wave produced by the sledgehammer to reach the geophone or to pass between geophones.

At our study site in Texas, the seismic component of an elephant rumble was much shorter than the 56-foot airborne wave. It traveled at about 820 feet per second, which amounts to a 41-foot wave. Given the difference in wavelength between the two signals of about 16 feet, and the difference in velocity (airborne sounds travel at 1,083 feet per second), if an elephant were really tuned into the signals, it could detect the difference in time of arrival between the two different wave types, acoustic and seismic, and estimate the distance of the vocalizing elephant. On top of this, the fact that the seismic wave could be either shorter or longer than the one traveling in the air could help to localize extremely long waves. This is true for two reasons: first, simply because it is a different length; and second, because the distance between the elephant's front and back feet is much greater than the distance between its two ears, providing a greater distance to determine differences in the angle or position along these long sinusoidal waves as they arrive and are sensed by the foot as opposed to being detected by the ear. Alternatively, the greater distance between sensors provides a longer distance between the time of arrival at one sensor as compared with the other, which could also facilitate localization.

The capacity to detect both air and ground signals had the added benefit of potentially determining the distance of the signal source, assuming the airborne and groundborne waves had two separate velocities. This concept is akin to counting the time between seeing lightning and hearing thunder to estimate the proximity of a thunderstorm.

We returned triumphantly from our Texas safari, analyzed the data, scoured the literature for supporting references, and then quickly put together a paper to present our findings at several scientific conferences. The idea resonated with many colleagues, but there was some criticism from a reviewer familiar with the military literature that apparently showed that soldiers in Vietnam were not able to use wave detection to measure the approach of other soldiers from a distance. We countered this argument by suggesting that there was too much background noise from tanks and other military equipment to pick up human footfalls. The reviewer didn't buy our counterargument, so we had to conduct our own demonstration of long-distance footfall detection, which Byron called, "Operation Jumping Man," in the silence of his Nevada gold-mine site. OJM helped us overcome any further criticism as we were able to show that the signal from a 170 pound man jumping up and down was measurable above the background noise one kilometer away.

In the meantime, I tried to set up a site in a zoo where we could ask very specific questions about an elephant's ability to detect vibrations, using an elephant that could be trained to respond to various vibrations at lower and lower levels. I wanted to understand how far away an elephant might be able to detect seismic waves.

Working with captive elephants was a difficult adjustment, since my only experience with them had been in the wild. It took me a while to adapt to the constraints, the politics of zoo management, and then, of course, the conditions of captivity for elephants. Around the end of 1997 I met Colleen Kinzley, the general curator of the Oakland Zoo. When I visited the zoo, I was delighted to see the elephants' rich characters and how caring and conscientious the keepers were. We scheduled a date to work out an experimental design with their most amenable elephant, Donna.

10

IVORY GHOSTS

It does not require many words to speak the truth.

—CHIEF JOSEPH, leader of the Nez Perce (1732–94)

IT WASN'T UNTIL 1998, three years after I had gone back to the United States to work on my elephant seismic communication theory, that I was able to return to the Caprivi. I had gained the attention of Rotary International for my work on elephants and farmers in the Caprivi, and they supported my return to Namibia as a vocational scholar to assess the results of my conflict mitigation project and to put in place more elephant crop-raiding deterrents.

A lot had transpired since we had left the region. A veterinary fence had gone up along the Botswana/Namibia border, funded by the European Union, to protect cattle from contracting foot and mouth disease from wildlife, but it cut off the migration routes of one of the last migratory populations of elephants left in Africa. Jo Tagg took us down to the fence to see its devastating repercussions. A kudu carcass hung impaled on the fence, most likely from exhausted attempts at escape.

Fortunately, the following year, NGOs in Botswana obtained Tim's elephant satellite tracking data to prove that the Okavango delta was an important wet-season habitat for elephants in the Caprivi. Armed with this data, they were able to convince the government of Botswana to take down 12 miles of fencing on both sides, allowing free passage of elephants and other migratory species along the main migratory corridors of the Kwando and Kavango Rivers into the delta.

The troubled politics of the Caprivi continued to escalate, and the day we drove through the Ngoma border from Botswana, there was a trail of secessionists leaving for Botswana. Apparently, some Caprivians were tired of being controlled by politicians in Windhoek who they felt were not acting in their interests.

We had come from Kruger National Park, where Tim was finishing his dissertation on tuberculosis in the Cape buffalo and I was writing up our elephant/human conflict study. When we left Kruger, we drove across the Limpopo River and up through Zimbabwe, where we met up with my dissertation adviser and her husband at Elephant Camp, just north of Victoria Falls. I had arranged to do some seismic playback studies on a group of eight trained and semicaptive elephants at this site.

The results from these early playback trials were promising enough to encourage us to continue them in both captive and wild settings. We knew how difficult it was going to be to isolate vibrations within the ground in the wild. But in some of our more successful attempts, one young female elephant named Miss Ellie got so upset by one of the seismic playbacks that she bent down and bit the ground, an action that is seen in the wild only under extreme agitation. We did not repeat those experiments with her, since she was obviously very sensitive and we did not wish to cause her undue stress.

When we finished with these studies, we said our good-byes and Tim and I headed back to the Caprivi. Tensions were high in Katima due to the secession. There were stories of mistaken identities and accidental shootings, of Bushmen running scared to relatives in Botswana after being threatened by Namibian soldiers, including the Bushman chief. Various political parties condemned the activities of the alleged members of the Caprivi

Liberation Army, calling the bloodshed "undesirable and contemptible."

I met up with Janet Matota to discuss the progress of her role as resource monitor, the status of elephant conflicts, and the history of the successes and failures of the electric fence at Lianshulu. Due to the lack of motivated people to keep it running, the fence lay in ruins after several successful years of deterring elephants.

Janet was in her usual good spirits but more distant than she had been when we were working together. A lot had transpired since then, and the weight of it was evident in how she carried herself. Furthermore, the added responsibility of having to maintain a pickup for her job was becoming difficult to manage.

She summarized things first in terms of all the people who had recently died, either along the Golden Highway or of AIDS. We began by mourning a ministry ranger and close friend who died in an accident on the Golden Highway. Then there was the list of AIDS victims. Janet always began her letters and our greetings this way, as though mortality were a measure of time. Things were getting so bad in the region that the post offices were now offering special prices on coffins with plastic flower wreaths, the post office serving as a makeshift repository, delivering both mail and loved ones to their respective destinations. I only hoped that they were more successful in delivering coffins than they had been the mail.

We had gone to one of these outpost offices together one day to use the only phone line outside Katima. This was a huge sign of progress for the region and meant that I could bring in my laptop and connect it to the phone line to collect e-mail. This advance came several years after our long residence, when we had been stuck in the bush with no communications. It caused quite a stir for everyone waiting in line to use the phone to see me plug the cord into my portable computer. Rows of wide eyes watched the mysterious bar get larger and larger across the bottom of my screen, representing the inbox filling up. When I was finished, I simply plugged the wire back into the phone and gave the poor, confused postal worker his fee for the amount of time I had spent with the phone line disconnected from the phone. There was no dial tone, no words spoken, nothing but a moving blue bar. Heads shook, and the whispered words "white man's magic" could be made out.

The Katima doctor was finally coming under heat for his ivory dealings. He had been trading across the border, flying a plane into Angola with a red cross on it under the guise of delivering medical supplies. Some of his ivory-smuggling missions even operated via the landing strip at a local lodge late in the night without the knowledge of the managers.

Janet told me that a renowned poacher in the region had escaped from prison and finally met his end in a skirmish with the Botswanan Defense Force (BDF) across the border, after an untold number of elephants and other wildlife had died by his hand. She was told that he had gone across the border bearing an AK-47 and the BDF had had enough of him and did him in. Apparently, they shot him in his elbows and knees, then dragged him behind a vehicle before finally shooting him in the head. The message quickly reached other poachers. At one time, he had a big WWF sticker on the door of his kraal, the familiar black-and-white panda symbol, ironic proof of his "interest" in wildlife.

I asked Janet if the poacher's brother was still selling body parts to the witch doctor. She laughed and nodded. In our long rides together, she told me all the local gossip with a youthful glint in her eye. One day, she explained that there was a whole mail-order industry based in South Africa for medicines to become smart or rich. Many people were buying these products, but apparently they were dangerous if not taken as directed.

"I was told that there was a man in Katima who wanted to become rich, so he took the medicine. The directions said that after you take the medicine, you must sleep with your sister or your mother. And you see, it is not in our custom to sleep with a sister or mother, so he was very afraid to do that thing.

So he did not follow the directions, and he became very ill. In fact, you used to see him walking around Katima with an ax. He was not well. Then one day, he decided to look for his mother so that he could follow the directions and get better. His family hid his mother away so that he could not find her. He walked around Katima for a year. Finally, he found his sister and raped her."

"He did?"

"Yes, and I am told that he is normal now."

"But did he get rich?"

Janet explained that people believed that he still had what she called "a kind of small gorilla," putting her hands 6 inches apart, meaning a bush baby or night ape, a very tiny primate. Part of the medicine included a bush baby that was kept in a box. It was supposed to go out and get money and put it in the box, collecting so much that the person would get very rich. And if the medicine was taken correctly, the bush baby would stick around and collect money forever. When I asked what the bush baby was fed, Janet shrugged and said that it was thought to be "a sort of magic, that it did not need food."

She had told me that the poacher's brother was getting richer by the day because apparently he also had one of these small gorillas.

"But isn't that because he sells body parts?"

Janet looked at me, not realizing that I knew about the body parts. "You see, Catleen, he sells to the witch doctor. He makes very much money. In fact, one day he sent some kids to Sauzou to hide in the mealie fields and wait for a small boy to catch."

"He sends out kids to do his dirty work?"

"Yep. But that day the kids were caught. They told everything. Then when he caught them, he took them out to the bush and beat them."

"I heard he was caught with a trunk full of dried human testicles. Where does he get them all from?"

"He hears when there has been a murder or fight, or else he

just sends out people to retrieve some." Janet smiled at my look of disgust.

"I guess we'd better watch out for the 'retrievers'!"

After my meetings with Janet and some of the game guards and other IRDNC staff in the region concerning plans to reinstall the fence, Tim and I said our good-byes and planned to head to Etosha to visit other friends, then to Windhoek to buy supplies for the fence before going to Capetown.

As we left, people boasted about the quality of the partially paved Golden Highway and how traversing the Caprivi was now much easier. Tim looked forward to the experience, but I was anxious. I had had an accident along this stretch of road in its early, treacherous days, and returning to the Caprivi after so many years brought back a wave of emotions, my grief refusing to go away. I relived the scene countless times over the years through flashbacks; the wound would simply not heal.

We drove along the infinite stretch of smooth tarmac before reaching—there they were—the exclamation points in the middle of the road to mark the slippery sections, but to me they served to draw attention to the graves that piled up on the clay patches of the wet-season fairway. I couldn't help but wonder how many lives were claimed prior to the posting of exclamation points. Then I found myself reliving that day.

SITTING IN MY TRUCK THAT DAY, I had waited at the gate to the West Caprivi Game Park while a soldier entered my license and registration number into a logbook. There were several soldiers standing in front of the pickup armed with R1's. The policeman came over to me, handing me the register. I filled in the date, my name, and a host of irrelevant details that nobody ever bothered to check.

I looked at the deep purple sky, the air washed clean from a night of torrential rain. The road ahead looked like a sheet of ice. Mine would be the first tracks to traverse the snow-white road since the early morning downpour. A soldier held the gate while many people piled into the back of my truck. I asked if he would open the gate, signaling that the back of my truck was full by placing my hand on top of my fist and that I didn't want to take any more passengers. The soldier looked at me gruffly and held the gate closed until the back of my truck was loaded further. I wasn't sure how many passengers there were. Because of the lack of public transport in the region, the accepted policy was to take on passengers whenever possible. But the number always surpassed my comfort level.

I looked down and brushed mud off my shirt from having to change a flat tire that morning. I was coming from Mahango Game Reserve, where I had spent the night in an empty army cabin. The rain's echo on the corrugated iron still rang in my head, and I wasn't feeling very well after the long, restless night alone. I had stopped there only because I hadn't left Etosha early enough to make it home before having to drive the West Caprivi road on a rainy night. It was too dangerous.

When the soldiers finally stepped away from the truck and allowed me to proceed, I drove through the gate and onto the slick road, gripping the steering wheel to negotiate the slippery patches of clay forming shallow pools in the gravel road. My schedule was shot. When the road was good, it took two hours to cross the 125-mile stretch. Today it could take at least three, even four. And the Mayeye khuta meeting that Matthew had arranged for me was an additional hour's drive off the Golden Highway. We were hoping to get approval to start the women's resource monitoring program in that region.

The truck quickly became caked with white, sandy chunks of clay. The windshield wipers struggled to clear an oily mixture spewing up from the tire wells. There were patches of showers, but the heavy storm kept pace just ahead, washing out the road and making the clay lakes like sheets of ice with no tracks to follow. In the rear-view mirror, a shapeless, fearful clump of human clay filled my view. I felt lonely. Guilty.

All my instruments sat piled up in the passenger seat—it was my responsibility to protect the ministry's equipment. Did they really expect me to choose instruments over life? The government hoped to rise above its perpetually overextended condition by taking on parties from the outside whose assets could remain neutral. But how neutral could I remain while the people around me had nothing? These "responsibilities" often made me feel subhuman. And then I felt even lower for not having the strength to "interpret" the rules, as did the region's local hero, Simon Mayes.

Simon held the equivalent of Matthew's position in the West Caprivi, and they lived together at Susuwe. After being posted in the Caprivi with the SADF, he stayed on as a local conservation official within the Ministry of Environment and Tourism. He later left to work with the IRDNC. Simon was renowned for his acts of kindness, some simple, some great, but all symbolic of his respect for life. So whenever I tried to uphold order, an attempt to protect the ministry's interests, the people would invariably exclaim the magic words, "But Mr. Simon would do it!" And with that, I did whatever they asked me to do, regardless of the potential risk to a ministry asset.

I saw a clay patch approaching and I began to slow down. This time, the clay extended into the gravel, causing the tires to slip sooner than I had anticipated. I was going too fast to control the weight of the truck, which began to slide sideways. I took my foot off the accelerator, wheel against the skid. I might as well have been on ice. *No brakes, no brakes, let it ride.* I thought the truck would spin, that I could regain control when it lost momentum.

Instead, the left wheels rose from the weight of the human mass forced against the right side of the vehicle. The ground approached very slowly. Then a blur. I held onto the steering wheel, pulling legs in, the door now forced open from my weight. Upside down. Explosion. Oh my God. Reverse. Rewind. Those people. The new truck.

The truck rolled once and landed upright. Above the smashed windshield, the crooked wipers convulsed in midair. I sat shocked in a bath of crumpled glass, mesmerized by the tortured wipers in their frenzy.

Turn them off!

The sickening silence throbbed in my head. And then reality hit hard. All those people. I undid my seat belt. The door clung to its hinges like a piece of crumpled foil.

I got out, feeling eerily suspended above the scene. Scanning the road, I quickly assessed the status of seven scattered forms. There was one still in the back of the truck, heaving over the side. He had gotten stuck on the roll bar. I tried to lift him off the truck, but he was too heavy. The boy's condition seemed urgent; he was unable to move. The rest of the passengers were standing still, trying to piece together what had happened.

I grabbed a large figure still dazed and holding his head. "Help me get him down!" I shook the man and got him to help lift the boy off the truck. The boy fell into a pool of white slippery water, flat on his back, desperate to breathe. I tried to soothe him, to get him to breathe more smoothly. His fearful eyes were focused on a distant looming force.

I called a man in a military uniform over to help us move him onto higher ground and then quickly walked over to three young girls who stood in the middle of the road, shocked and whimpering. One had a huge lump on her forehead, as if someone had opened her scalp and inserted a golf ball. Another had a broken arm. The other stood there wide-eyed, recounting. I gently hugged each of them in consolation and instructed a small boy to help make a sling for the girl with the broken arm before returning to the quivering white form.

The boy now managed to half fill his lungs with quick gasping breaths that took all his remaining energy. I wished that I could open my watch, take the second hand, and wind it back three times. Just three. It seemed easy enough. Easier than willing myself to vanish. In stilted Afrikaans, I tried to make a plan with the two men. I looked down the long white corridor of road as it vanished in the distance. An echo enveloped the empty road, reverberating as I turned my head in either direction. No sign of life.

"Hoe ver is Omega?" I asked.

A small boy signaled in Afrikaans that we should drive the truck to Omega, a village 6 miles away. Drive the truck. I looked at the crumpled cab. Of course. I walked to the cab and started the engine. It started, clucked, and roared to life. I left the engine running.

The body on the ground started to convulse. The smaller boy approached me in an ominous tone. "Ek dink die boy is a bietjie dood," he exclaimed, gauging my eyes for a reaction. "Die boy is a bietjie dood!" he cried pointing over to the cold body. "I think the boy is a little dead."

I walked over to the prostrate figure. No. I panicked. This is not really happening. No. It can't be. He is just unconscious. He's not breathing. I positioned the boy's head to give mouth to mouth. It was evident that the others had never seen or heard of such a procedure.

Holding his nose, chin down, I pursed my lips and breathed. Nothing. I breathed again. Nothing. No pulse. I pumped his chest three times and breathed again. The two men and the small boy looked at me as if I were trying to beat life back into him. This time I forgot to plug his nose and blood spurted from his nostrils and trickled down his mouth. The soldier urged me away, the blood a definitive sign of death. My eyes froze in horror on the bloody nose as the soldier took my arm and led me away.

"Come, don't worry," said the soldier. "You are helping us." I was surprised and strangely comforted that he spoke English. He tried to pull me away from the dead boy as I sat paralyzed with disbelief. All those unseen years snuffed out.

"You gave us a lift," the soldier said gently. "You were helping us. It is not your fault."

I sat in the puddle facing the boy and rinsed my mouth out. Trembling hands in milky water. Grit and white clay in my mouth. White. White. White. This isn't real. Why can't I be dreaming?

I covered my face with my white hands and dragged them down my face. I stared at everyone through my streaming eyes. They all looked at the ground, away from the corpse.

"Come, it was an accident. You are helping us." The soldier led me toward the clacking truck.

Why is no one reacting?

Here, even a wounded leg would eventually give way to a death certificate. Acceptance was part of survival.

"When I was in the SADF, a helicopter would have arrived," reflected the soldier. "Why can't this government do such a thing?"

I looked down the empty corridor of road. Tears blurred the scene as I watched the survivors fill the back of the truck again. They couldn't just toss the dead body on the back. I quickly unrolled the bedroll that I stored in the cab. As I spread it out in the back, the girls very gratefully climbed aboard. I tried to signal to the soldier that I wanted the body to be laid on the bed. He calmly indicated that there was plenty of room for all, including the dead. The dead boy was placed on the edge of the bed, the living giving him a wide berth.

We puttered into Omega, a familiar sight, with no windshield and the door held shut by hand. There were so many accidents on that section of road that this was a regular sight, marking the height of the wet season. I pulled up in front of the police station. Ghostly forms lined up to collect the death toll. Silent eyes looked on as the passengers spilled from the back.

Having been here a few months before to follow up on a poaching case, I remembered that the police had been very friendly. I recognized some of the officers as I stood in silence among the crowd with eyes down, waiting.

Then I heard the first cry of mourning, from the boy's younger brother. He screamed and jumped at the body with hysterical sobs as others pulled him away. I stood next to the vehicle and allowed the crowd to absorb me. I waited for the anger as the wailing heightened. There was none. Pity. Shaking of heads. Clucking. Looks of sympathy. I wept openly, standing in the crowd for a long time, hoping to convey my regret.

By standing with them, strangely, they allowed me to be part of them, sharing and uniting with them in their grief. It was an uncanny union for which I felt grateful but still shattered.

A while later, a hand on my shoulder guided me into the police office. A friendly voice asked from the darkness, "Where do you live?" It took a moment to adjust to the darkness. The first thing to come into focus was the contents of my cab sitting on the floor next to the desk. I recognized the random collection of garbage from my road trip. The Cadbury "snacker" from yesterday's lunch, a granola bar wrapper from breakfast, some empty "AfriCan" giant-sized Pepsi cans, a trashy fashion magazine I had picked up in Tsumeb, Palgraves tree book, and my equipment cases sitting next to the blood-spattered bedroll. I felt so exposed. Why did they bother to save the garbage? I was strangely embarrassed about the fashion magazine, then profoundly ashamed that I could feel embarrassed about such a trifle, given the situation.

"Susuwe Ranger Station." I could now see the policeman whom I had met last month. Patrick, the stubby, gentle man with piercing eyes. But this time his eyes revealed a little too much interest. He asked that I return the following week to visit the scene of the accident and fill out a police report. He touched my wrist. Vulnerable. Weak. Needing consolation. I hoped I wouldn't have to return alone.

"Is there anyone who can pick you up?"

The police van, the only extra vehicle, acted as a makeshift ambulance to take the wounded passengers and the dead boy to Andara, the Catholic mission about one hour to the east. I was stranded.

"Is there a phone I could use?"

Patrick turned to one of his underlings. "Moses, take this young woman over to Hamukoto's office. The phone over there should be working."

"Yes, sir."

"Are you all right?"

"Yes. Thank you." My voice cracked.

Moses led me to a large, empty cement building, and I followed him through a maze of hallways, past empty offices with no windows, until we reached an office with a heavy wooden door. Moses knocked on the door, and a muffled voice could be heard from inside. Moses walked in.

There was a large man sitting behind an immense desk, his belly barely contained within his khaki safari suit, hands clutching the edges. The only thing on top of the desk was a black telephone.

"The madame is in need of the phone."

Cold dark eyes stared back at us.

"There was an accident."

The large man pointed at the chair in front of the desk and picked up the receiver. He pressed down on the button several times to get the operator.

"Where to?"

I sat down. "Ministry of Nature Conservation, Katima."

The man's eyes narrowed as he stared at my face.

"Yes. Get me Katima 52," he ordered and put down the phone, placing his good hand over the one with a strange, bloated deformity. I tried to maintain a pleasant expression, but the man unnerved me with his stare.

The phone rang.

"Yes?" He handed me the phone. "You are through."

"Hello?"

"Moré?"

"Tokkie?"

"Jah?"

"This is Caitlin. Is Jo there?"

"Hold on."

"Hello, Caitlin. How's it?"

"Jo?" I mumbled weakly. Hearing his voice set me into another fit of panic.

"Are you well?"

"I had an accident." Tears streamed down my face as the man stared at me with his hands folded over his belly.

"Really, hey? Are you all right?"

"No." I hesitated. "Jo, someone died."

"Jesus. Are you okay?"

"Yes."

"Where are you?"

"Omega."

"Will the vehicle run?"

"Yes."

"I am coming just now. I'll take Charles along to drive it back."

"No, it's okay, Jo. I can drive it."

"No, Caitlin. Wait there and I'll be there just now."

"But really, Jo. I'm fine."

"It may start to rain again. Just stay put. I'll be there now."

The phone went dead. I set the phone back on the cradle and stared at the floor. I knew I would have a long wait.

"I will have my wife clean you up. You can wait at my home."

Stunned by this unexpected kindness, I graciously accepted the man's offer and followed him out to his car.

"This is very kind of you."

"One must look out for those in need."

"Thank you." I hesitated, not knowing the man's name, yet feeling it an awkward time for introductions.

The man kept his eyes on the road. "Homukoto."

"Thank you, Mr. Homukoto."

He drove past several Bushman settlements within Omega.

The sandy yards of run-down barracks were littered with rusted tin cans and old plastic bags.

"What do you do in Omega?" I managed, trying desperately to summon my conversational skills.

"I am in charge of agricultural development in this area."

"Oh." I remembered my last interaction with the "development crew" and felt a little uncomfortable. When Tim and I came to track down the poached elephant, a man from this group attacked the young Bushman boy who came along to help us. He choked him through the window of the truck, the whites of his eyes showing and teeth bared like an animal's. Tim put his foot to the pedal the moment the boy's throat was relinquished, and the poor boy apologized profusely as if we were somehow inconvenienced by this slight delay.

This project was brought into the area after independence as a way of bringing jobs to the former SWAPO (South West Africa People's Organization) freedom fighters by setting up agricultural stations. During the war, the Bushmen were brought to the Caprivi from Angola and Botswana to work as trackers for the enemy, the SADF (South African Defense Force). Thus, they were resented by SWAPO members.

Mr. Homukoto pulled up to a large one-story cement home, the only thing that looked loved in the whole town. Vibrant bougainvillea draped over the front of the house and a lush, green carpet covered the ground. A large woman emerged from the house and greeted us, taking me by the hands and leading me inside. How did she know?

"I will send Mr. Tagg here when he comes," Homukoto called out as he started his bakkie.

I nodded back at him, wondering how he knew Jo's last name. I suddenly had a flash of seeing this man before in Jo's office. What was he doing there? Was it something to do with ivory? Hopefully something to do with agriculture, rather. I hoped that my presence here was not going to cause Jo any unnecessary political complications.

The house was furnished with Western furniture, Afrikaans wall hangings, and knickknacks warming the hearth. Mrs. Homukoto led me into a fully plumbed bathroom and closed the door. The movement of a ghostly white figure startled me in the mirror. I hadn't realized that my hair, face, and arms were completely covered in white clay, except for the tear marks forming ribbons of flesh on my cheeks. It was my pointless attempt to wash out my mouth, the clay drying into a sad clown with pale lips and wide eyes. I looked at a forgotten me, the me before Africa. Something about the expression in my eyes, innocent, seeking protection. Who had I become? I hadn't seen that face in a very long time. I stared into my eyes. What am I doing here?

Somehow Africa isn't a place for mirrors. We had a broken sliver of a mirror in the bathroom with which Tim used to shave, and I used it on the rare special occasion when I wore eyeliner. I couldn't remember the last time I'd seen my whole face.

I started to cry again. I stared at my reflection for a long time, watching the tears flush the chalky banks of my cheeks. Maybe I should give up. What am I trying to prove? The only thing I had accomplished was angering a lot of farmers. I finally reached for the tap. There was no water. The tub next to the sink was full of water. I didn't know what to do. I didn't have any other clothes or a towel. I stared at the full tub. Slowly I started scooping some water from the tub onto my arm.

There was a knock at the door. Mrs. Homukoto entered with a towel and bucket and put me into the shower stall next to the tub. She filled the bucket from the tub, helped me with my clothes, and started to sponge me off. "I filled the tub before the water ran out," she explained.

I stood speechless, embarrassed that I had soiled the woman's water supply. She hummed as she gently scrubbed me down. I surrendered my body, absorbing the kindness, absorbing the warmth of her tune, wishing it was forgiveness.

When the woman was finished, she led me into a bedroom and closed the door. I nodded to her in thanks, walked in, and sat stiffly on the edge of the bed, my head ringing. I was exhausted but wired. How could I lie down at a time like this? And with my dirty clothes. It was 2:00. It would be at least another hour before Jo would arrive, maybe longer depending on the weather. I fidgeted, fumbling with my hands, and then started pacing the room. I eventually resigned myself to lying flat on the bed, staring blindly at the ceiling, trying to comfort myself with thoughts of home.

I saw my father's face the last time I talked to him on the phone before I left for Africa. I could hear from his voice that he was choked up. He thought for sure that I was throwing my life away, that he would never see me again, his daughter, possessed by some strange demon calling her to her own demise in dark Africa. And now I wondered, was he right?

Two hours later, Jo arrived to pick me up, and within minutes we were making our way back to Katima. As he drove, Jo tried his best to console me. It was an accident. It could have happened to anyone. They knew that. His rational approach was an effective temporary Band-Aid. Denial, perhaps, but I needed something to cling to for just a little longer.

When we arrived at the office, the staff lined up in an impromptu funeral service, each shaking my hand in turn, offering their condolences. Somehow I didn't feel I deserved this kindness but tried to appear appreciative.

Charles had taken the truck for a joy ride and didn't show up at the office until after closing. When he pulled up, Kai stared at the crumpled death trap with me while Jo berated Charles. Kai gently offered to share the weight of my demon with stories of his near misses on these bad roads.

Jo drew a bath for me, picked some tiny flowers and put them next to my bed, and asked if I wanted a sleeping pill. I soon regretted saying no and suffered through a tortured night, replaying the scene over and over in my nightmares. The boy's distant and haunted eyes looked up at me through slits. I tried to shut them, but they kept reopening, staring at me desperately, sadly.

Tim arrived the next evening from Windhoek, and we mourned together quietly in Jo's spare bedroom. I tried to let go. The numbness of tragedy is pierced only by those who know you best.

11

THE WHEELS OF CONSERVATION

Man Charged for "Raping" Corpse

The bizarre alleged rape of a woman shortly after she was murdered by her husband—after he apparently caught her making love to another man—has resulted in a Katima Mulilo resident being charged with necrophilia.

In the frenzied attack on Saturday evening by an enraged husband the woman, Asia Jowel, and her alleged lover, who has yet to be identified, were beaten with fists and clubbed to death with a stick.

The man who appeared before the Katima Mulilo Magistrate charged with the twin murders has been named as David Kaunyade (32), while the man who was caught raping the corpse of the woman has been identified as James Basson (38).

Kaunyade and some Policemen caught Basson, who was reportedly "heavily intoxicated" and passed out after being shoved off the corpse.

—CHRISPIN INAMBAO, *Namibian, September 29, 1999,*
Web posted at 10:40 a.m. GMT

AFTER CROSSING THE WEST CAPRIVI, we stopped in Rundu to get gas. We laughed as we passed the place that Jo called "a good place to get a steak and a broken nose!" When we got to a farm road just outside of Etosha, we remembered the time when I was to meet Tim after a long separation, and on my way there from the Caprivi, my gas filter got so clogged from an accumulation of dirty gas that I could no longer accelerate.

It was a Saturday and none of the service stations in Tsumeb had a mechanic on duty. The gas attendants shrugged at my feeble explanations, because to them, if the truck could make it into the station and I could drive it out, there was no problem. I forged ahead, hoping at least to make it into the park before the gates closed at sunset. I was nearing the park entrance when I realized it was hopeless. I knew I'd be safer spending the night on a small farm road rather than remain on the main road to Oshakati.

I pulled onto a farm road and made it to the edge of a small holding. Since it was a weekend, the owners were away, but I wanted to let the farm laborers know what I was up to. Maybe out of pity but probably because he didn't completely understand my dismal Afrikaans, the head laborer told me to talk with the owner. He led me into the farmer's house and called the farmer on the phone. I explained my situation to the very friendly German farmer, who welcomed me to make myself at home in his driveway, where I looked forward to rolling out my bedroll and sleeping under the stars.

I thought I'd try to call the research station at Okaukuejo, knowing full well that the receptionist was off for the weekend and no one would answer. Luckily, Tim was sitting right there working on his data and answered. Tim decided to race through the park, beg the gatekeeper to let him out after hours, and meet me at my makeshift campsite on the German's farm, where we shared the night with meteor showers and contemplated the largest meteorite, the Hoba, which fell just outside the next town more than eighty thousand years ago, all 30 tons of it. It wasn't far from the farm with the dinosaur footprints looking like great ostrich feet walking into the rocky distance.

The next morning, Tim explained the symptoms of my truck to the head laborer in Afrikaans, who promptly took a bicycle pump, attached it to the gas filter, pumped out the junk, and the truck was as good as new. We waved a cheery good-bye, and the man was happy to find himself fifty Namibian dollars richer.

After a visit to Etosha, we passed through the thornbush savannah of the central plateau to get to Windhoek, game fences lining the road most of the way, passing the Waterberg plateau on the left and the Skeleton Coast off on the right beyond a great expanse of lowland savannah spreading out as far as the eye could see. The only topography in the foreground outside Otjiwarongo were the two Omatako peaks that looked like two great teats rising up at the top of a long nubile belly of some ancient Namibian queen. This land was connected to Brazil 135 million years ago before the breakup of Gondwanaland. Namibia still teemed with gems, and diamonds once littered the sands of the coastal Namib Desert, fed by the cold nutrient-rich Benguela Stream.

After a brief stop in Windhoek, we headed through the open rocky desert, crossed the Orange River, and traveled through the dry, succulent Karoo until we finally descended into the fertile Cape. When we arrived in Capetown, I met up with my local

Rotary sponsor and said good-bye to Tim, who had to return to the States. Then an old friend flew in who would accompany me back to Namibia. I planned to drive through the Richtersveld and up through the Namib Desert before returning to Windhoek. I would finalize the details of my Rotary project there before heading back to the Caprivi to build the fence.

With knuckles white on the wheel, I braved the stark savannah of the Richtersveld. The butter tree dangled its red trumpets from an overstuffed, jaundiced limb. A pair of spotted eagle owls sought a sliver of shade on a ledge of a dolerite outcrop. Their angry stare heightened my sense of being a stranger in a forgotten land.

From the Richtersveld, we headed for the Namib. On my very first visit, I had stood in awe overlooking the expanse of dunes at Dead Pan. I was a different person then, unable to decide if what lay before me was rock or sea. With my feet submerged in the chilly apricot-colored dunes, I felt that life was uncomplicated. I was now back to feel the dunes with a new perspective, the rhythm of Africa part of me now.

We packed up and headed north to the Caprivi. I felt ready to return. After many more meetings with farmers, debates continued over where to move the boundary of the new fence. Time was running short, but at least we managed to get a good start on rebuilding the electric fence around Lianshulu village, with Janet, Loveness, and her husband, Beaven, the head game guard for the region, playing a pivotal role in keeping the community united and the farmers interested.

As we argued about the fence, the end of my Rotary fellowship was fast approaching. I brought all the supplies to the game guard office, and Beaven assured me that they would complete the task with the farmers after I had left. After that, I decided to take the opportunity to say good-bye to friends by attending a local ceremony for a conservancy that began as a suggestion in our first khuta meeting, when we were introduced to the Bukalo chief.

I stood at the Salambala Conservancy inception ceremony, elaborate for a local affair, with the schoolchildren from the Catholic mission lined up in their uniforms, singing songs for the guests, then some words from the local chief. He requested the help of the local NGOs to remove the last family refusing to leave the property of the conservancy, since their persistence made it impossible to comply with regulations.

I looked around the crowd at all the dedicated participants in this effort. Garth and Margie from the IRDNC were there, along with the head of WWF–US and colleagues from the Directorate of Environmental Affairs, all of whom could see the long road behind them—as well as the long road still ahead. Much had been accomplished in the few years since I left the Caprivi. There were now a number of emerging communal area conservancies, where elephants and other wildlife were at least welcome and under some form of protection. All of these areas could have been converted into cotton and cornfields had the vision for conservancies not taken hold. On top of this, people were starting to see a real financial return on their stewardship. As I listened to a final song about unity and gathering at the river's edge, I reflected on how this particular venture had begun.

ONE MORNING, we all met at Jo's office at 8:00 A.M. as planned. Kai, Matthew, and Shedrack, one of the rangers, were talking triumphantly in the reception area as we walked in. They were gloating over the latest victory in their ivory case against the town doctor. The room fell silent as we walked in, since they did not want to involve us in confidential matters. We greeted them and ducked our heads into Jo's office to escape the awkward situation.

"How's it!" Jo exclaimed, waving us in. "I am just waiting for a phone call, quickly. From head office." He pointed to the seven-tusk haul, an addition to the bust they had made the week before at the Piggery, an old slaughterhouse where all community action, both good and bad, took place. Same courier. Same ultimate suspect. "Then we can leave." He rubbed his hand down his tired face. "Come, sit down."

As we sat down, Jo told us that a man had fallen into a volcano in Hawaii the day before. He burst out laughing. "I love it. Heard it on the BBC this morning."

"Fell?" I laughed.

"Yup!" Jo gave a wild-eyed giggle. "I love it. Perfect, isn't it? Makes you feel that Hawaii isn't so tame after all. Far more alluring now with an element of danger."

"How did he fall in?" Tim asked.

"I don't know. Guess he stood too close!" He squealed with delight. "Keeps you on your toes." Laughing, he waved his hands. "You know, you could get eaten by a lion or trampled by an elephant in the Caprivi, but Hawaii could swallow you whole! No chance for retribution, no opportunity to stare down the challenge, nibble at fate with a small-caliber handgun. Nope, just one gobble. Mother Earth has no mercy! It's wonderful! I thought this was the only place that sent one into darkness!"

The phone rang. Jo jumped into serious mode and waved us out of the office.

Matthew, Kai, and Shedrack were now engaged in small talk and happily included us as we waited for Jo. When Jo emerged, everyone climbed into the land cruiser. With Jo at the wheel, we headed toward the eastern floodplain. Our meeting was scheduled for 9:00 A.M. With such a large agenda, we knew that it could last well into the afternoon.

It had rained heavily the night before, but it was easier to anticipate the slick clay in this area because the rains created large pans next to the road in the open mopane forest. Jo slowly navigated through the swampy mud, clay, and sand, with a steady stream of jokes, successfully distracting everyone from the journey. We made a cheerful entrance into the turnoff and drove through the well-kept village of modest thatched houses, some with courtyards adorned with papaya trees and euphorbia bushes.

There was a lot of clay in the eastern floodplain, giving the buildings a stronger, tidier feel. Fed up with the floods, the Masubia people came from Zambia decades earlier when the chief sought higher land for his people. The chief of the Mayeye people allowed them to occupy this land. Of the three main tribes in the East Caprivi, polarizations oscillated between the Mayeye, the Mfwe, and the Masubia at different times for different reasons.

Jo parked the vehicle under the shade of a silver terminalia. We all got out and stretched our legs. A few minutes passed, and a messenger came out to invite us into the khuta, where all the tribesmen and the chief were waiting to start the meeting. After the messenger showed the way to the front door, he stopped me at the entrance and signaled for me to wait until everyone else had entered. Then he led me around the back of the khuta, through the kitchen courtyard where the women were stirring black pots as big as buffalo.

A few chickens squawked as the messenger guided me through the back door. As he showed me in, he apologized. "Sorry, we blacks . . ." I smiled, nodding my understanding. Margie had warned me that I would be led in separately, through the back door. On the rare occasion that a woman was brought in front of the khuta, she was not allowed to enter through the front door.

"Musuhili." I made an effort to make eye contact with the women standing in a veil of smoke. They nodded and looked back into their pots.

I got to the door of the khuta when Shedrack, the last of the contingent, was kneeling at the door, clapping one hand on top of the other. He stood up, took a step, and knelt down again to clap his hands. He did this one more time before arriving at the table where the chief was sitting. All the headmen were sitting on the floor on either side of the table. The chief sat impassively, swishing flies with a gemsbok tail swatter mounted on a carved ivory handle.

I followed Shedrack's lead, trying to look as reverent as possible without tripping over my skirt. There were primary school–sized desks lined up against the wall facing the khuta. Jo wedged himself into one awkwardly, trying to ease the pain on his kneecaps by turning his legs in sideways. There was one desk open for me at the end of the line. I slipped behind it, aware of all the eyes on me, a white woman in a khuta meeting.

Jo was eager for the meeting to get started. He turned to Kai and then to Shedrack, signaling strategies with his stiff open hands. I looked at the chief in his double-breasted pin-striped suit, a stern look on his wrinkled face. The ingambella, second in command, was quite handsome with his mariner blue eyes and white hair. The headmen sat in a line wearing their finest yet tattered Western clothing.

The meeting began with a prayer, after which Jo thanked the chief for taking the time to meet with us. Then he presented the khuta with the agenda, introducing Shedrack Siloka as the new chief regional ranger. The ingambella nodded at Shedrack, and Jo continued introducing us. Some eyebrows were raised at the mention of problem elephants. Several members of the khuta looked me up and down curiously. There were a few snickers, as the community had become very cynical about the ministry's role in helping them with problem animals, particularly elephants.

Jo then presented Matthew's idea of starting a community game guard program in their area. He waited for a reaction, and all heads nodded for him to continue. He suggested that they revisit the suggestion of making a community nature reserve, an idea offered by the chief but never properly dealt with, explaining that new legislation was in the works that could be implemented to form a conservancy. Finally, he said that the issue of ivory needed to be addressed.

The chief shifted in his chair, a knee-jerk response. Jo continued firmly, "As elephants have been deemed 'specially protected game' through an international law body called the Convention on the International Trade in Endangered Species (CITES), no one is allowed to be in possession of elephant products. If they wish, they can apply in writing to the permanent secretary for special consideration." Jo eased up. "That is all we have brought to discuss today." There was silence for a moment as the khuta digested the proposed agenda. The chief and the ingambella whispered a few exchanges.

The chief spoke slowly, allowing the interpreter time to translate. First he looked at Tim and then at me. "I would like to hear what the elephant people will do." Then he turned to Matthew. "I have heard good things about your community game guards and would be very interested in discussing the project for this area." He then looked at Jo gravely. "I am angry about the ivory issue. The government is trying to ruin our traditions." The chief sat up and swished his fly swatter across both shoulders, allowing a suitable pause to put Jo and Kai on edge. "Nature Conservation and NGOs support tradition in other areas but not with ivory."

He started to list his grievances. "I am very angry that the Mfwe and the whites are working against the Masubia. The Mamili boundary in the Mayeye area was made without our support. When we wanted to give land to Nature Conservation, the offer was refused. Yet you make parks with the Mfwe. And the elephant problems on the eastern floodplain. People complain, but nothing gets done. I am upset that there is a person in the area who is being followed because he has ivory." He looked at Jo angrily. "A son should be allowed to bring the chief something he has found in the veld."

The chief paused, and the interpreter took a breath and waited for him to continue. "In 1989, that Salambala area was shown to Nature Conservation to be turned over to your protection, but you ignored our offer. The previous government was better because at least they allowed us to keep our ivory. The new government is bringing poverty to the Masubia. You must take these complaints to your superiors. The ivory must remain with the treasury of the tribe. These are treasures of the Bukalo throne. Tradition is quite different here than on the Western side. All khuta requests from this side are refused. And buffalo meat. Why is it that the other chief gets buffalo meat but I do not? We are worried that we have been discriminated."

Once the chief had aired all his grievances, he nodded his head, then the ingambella started in, and then each headman after him made a statement. When the khuta had finished, the chief spoke:

"I would like to hear what the elephant people are going to do."

Jo nodded at Tim to go ahead.

Tim explained that we had been hired to study elephants in the region partly because they were an important national asset but also because so little is known about this population, probably the last migratory population left in Africa. He explained that he would track them by putting satellite collars on them, receiving the information in Etosha National Park, monitoring movements to determine how much time they spend in the Caprivi and how much time they spend in other countries bordering the Caprivi.

Jo signaled Tim to let me continue. I explained that I would work with the ministry, farmers, and community, first to understand how extensive the elephant problems were and then to experiment with different solutions.

The chief responded, "If you want to see elephant damage, you can go to the floodplain. There you will see what the elephants have done."

The meeting went on for another two hours, and we all heaved a sigh of relief when it was adjourned. We headed back to Katima, discussing what was brought up and what needed to be done, especially investigating the Salambala Conservancy issue.

THE SALAMBALA CONSERVANCY had officially opened. The crowds were fed a celebratory stew from enthusiastic conservancy staff, all wearing matching commemorative T-shirts. New hope was in the air for a better relationship between the people and the wildlife.

The idea was that by placing control of decisions, wildlife management, and finances in local hands, the managing of conflict with wildlife would be seen as a community responsibility instead of the government's, elephants no longer being viewed as "Nature Conservation's cattle." This shift in responsibility would hopefully go a long way toward empowering the local communities and reducing conflict with elephants, as well as with other species such as lions. Seventy percent of the cases of cattle predation that I documented involved cattle that were left out in the park overnight rather than returned to their protected kraals. Progress, as measured by the advent of schools, also spelled doom for the cattle, as all the cattle herders were now in primary school.

Progress was also a double-edged sword for the farmer. I now understood why the local farmers no longer migrated to wet-season huts to protect their crops. With modernization came materialism and the cash that it took to get those possessions. With possessions came the need to guard them and consequently the beginning of a more sedentary life in the Caprivi, as farmers were reluctant to leave their homes for the time it took to grow their crops. The women would walk to their fields to protect them by daylight and then return home by nightfall.

There seemed to be only three choices for a peaceful coexistence between elephants and farmers: fence the elephants, fence the farms, or farm something that elephants don't eat. If the locals could eventually do away altogether with farming by enjoying revenues from ecotourism and other wildlife ventures, all the better for both man and elephant. But changing cultural practices was not going to happen overnight, and subsistence farming was going to remain an important way of life. It was probably necessary, too, to provide stability and independence during a time of transition toward sustainable wildlife management and utilization.

In the meantime, I was starting to see evidence that some of the local farmers felt the responsibility to resolve the conflict with elephants, investing more energy in learning how to maintain the electric fence around Lianshulu village, a project that in previous years they had felt was the responsibility of the government and then the game guards. Since I was starting to witness that shift, I was elated at the end of the Salambala ceremony, feeling that I was actually playing a role in attaining a larger conservation goal. I was handed a plate of fatty goat stew and relished it.

12

DONNA THE DANCER

The Hindus believed that eight giant elephants gathered
in a circle and held the earth on their heads.
Earthquakes were caused when one of eight elephants,
out of weariness, lowered its head and shook it.

—University of Memphis, Center for Earthquake Research and
Information (www.ceri.memphis.edu/public/myths)

BEFORE SHE WAS REBORN IN THE arid desert of Etosha pan in 1980, Donna had been a dancer in her former life. So an animal psychic had informed Colleen after meeting this contented elephant. Donna had immigrated to the United States along with many other two- and three-year-old orphans that had lost their families in the Etosha cull of 1983, a one-time harvest that park managers felt was necessary at the time in order to keep the population to a sustainable level. Donna now resides at the Oakland Zoo, and Colleen Kinzley, before becoming general curator, was her trainer for many years.

Colleen and I both thought it fitting that Donna had been a dancer. She had such poise and grace when standing on our force plate, sometimes balancing on tiptoes to reach a target that let us know she felt a vibration through her feet. And then there was the demi-plié in fourth position, weight on left foot placed on top of the shaker. It was those times when she looked most like a ballerina, and that position seemed to provide her with the greatest sensitivity to the vibrations entering through her graceful, if not dainty, legs.

There were other positions that were tango-like, determined, and controlled, feet planted on the ground, trunk out like an extended arm, then right foot swinging all the way across the front of her chest to reach the target that indicated "no." There were also the slow and thoughtful times, perhaps a little uncertain, when trunk tips wavered as she wafted across the plate. I saw her then in long, flowing taffeta, shifting her weight and trunk with a fluid motion, as in a waltz.

Donna took pride in her remarkably focused work. She seemed to view the experiments as a game, and each time she won, she was rewarded with treats, which came in such abundance that she would start salivating whenever she saw us coming with tables, chairs, equipment cases, and shovels. She once spontaneously urinated and defecated when she saw me, which I might have interpreted as a greeting if I didn't already know how thrilled she became, associating me with treats.

If the other elephants knew how many treats Donna was given during our experiments, there would have likely been an all-out riot at the elephant barn. It seemed only a matter of time before M'Dunda, Lisa, and Osh would find out what was really going on out there at the edge of the bull yard. Perhaps they, too, would join in the study one day, although I was a little leery of Lisa because she liked to throw things—rocks, logs, whatever she could get a hold of, including Colleen, who she threw against a wall one day, crushing her hand.

Osh once punched me with his trunk, so I never knew how close I could get to him and still be safe. He pulled it in like an accordion and then let me have it with a straight shot to the shoulder. I couldn't believe it. He was taught this playful gesture to spar with his trainers, and as such, it didn't hurt, but it sure did surprise me. Then he sort of rocked back and forth as if to say, "Come on, right back at me. Show me what you've got." I was so surprised that I couldn't think of a recovery. I probably lost a lot of ground with him that day. I was sure he'd be good at the experiment, though his penchant for play and his mischief made me reluctant to give him access to the thousands of dollars worth of electronics involved.

We began these experiments at the Oakland Zoo late in 2001. Donna was trained to lift her foot by pairing the command for foot lift with an audible elephant bull rumble played back through a low-frequency speaker. Lifting a foot was already part of an elephant's care routine, which allowed a trainer to scrape the dense cuticle off the bottom of the foot down to a smooth surface so that the natural deep cracks would not get infected from the unnatural cement surfaces of captivity. Within a single day, Donna easily transferred the command to lift her foot from a verbal command and a target stick pointing at her foot to the audible bull rumble.

Then we moved on to playing the rumble through a set of shakers, which were small sealed land mine–looking devices designed to mount under the car seat to shake to the beat of the car stereo. The shakers were buried in the ground about 20 yards away from where Donna would lift her foot in response to detecting the vibration through the ground. We experimented with making the trainer blind to whether there was a signal in order to eliminate the possibility that a sympathetic trainer, wanting the elephant to succeed, might unwittingly influence the outcome by an encouraging look or an unconscious gesture.

Finally, we switched from using Donna's foot to indicate a response to using the trunk so that she could keep both feet on the ground, focused on the vibration. In this scenario, Donna indicated whether she felt a signal by touching a square target with her trunk to her right for "yes" and a triangular target to her left for "no." We modified the setup by mounting a shaker under an aluminum plate where Donna stood and attaching a device that would tell us exactly the level of the signal delivered to Donna's feet, while minimizing the problem of the surrounding background noise in the ground.

Tellingly, Donna remembered her part in all this much better than Colleen and I did. We had developed different techniques of cueing each other when Donna got it right or whether we were ready to continue, but inevitably we made mistakes and were constantly reminding ourselves of the protocol, every once in a while messing up, much to Donna's dismay.

To let Colleen know that the experiment was ready, I would say, "Ready," and Colleen would stop feeding Donna and put her arms to her sides. When Donna had finished with her treat, she would lay her trunk over the "start" target, signaling that she was ready to begin the next experiment. Then Colleen would say, "Go ahead," when she saw that Donna was properly cued. At that point, I would either press "enter" on my computer to deliver a warble tone or I would not press the button if no signal was to be delivered.

If Donna responded correctly, I would say, "Yes," then Colleen would ply Donna with ample verbal praise and three treats in sequence, one treat at a time. "Good girl, Donna! Good girl, Donna D. That's a good girl." In the time it took for the three treats to be dispensed, I was able to recue the experiment and then announce that I was ready once again. When Donna answered incorrectly, I counted the time it would take to receive three treats and then started the experiment again to keep the timing consistent between presenting or not presenting a signal.

Early on in our experiment, I was hoping to wean Donna off receiving treats when she responded incorrectly. It was clear that verbal praise was an important part of the reward but not nearly as important as the treats. Colleen had been giving her smaller treats, negative verbal reinforcement, and negative body language by turning her back on Donna when Donna was incorrect. But Colleen felt that Donna needed a constant source of treats to keep her interested in participating in the experiment, since she was not tethered or in any way forced to participate. However, I was concerned about potential criticisms of our study should a reviewer feel that our results were invalid if we were providing rewards for both positive and negative responses.

When we tried to remove the small treats, it became immediately clear that the experiment would not work without them: Donna's tendency to demolish quickly kicked in. We decided that we had no choice but to use the relative reward, both verbally and with a treat, as a measure of success.

One day, Colleen was interspersing alfalfa cubes with slices of bread when Donna started to reject the cubes and turned out her trunk for more bread. We thought we might exploit her bias to our advantage by giving the least preferred rewards when Donna responded incorrectly. Donna was not happy with the new pattern, but it led us in the right direction, and we eventually succeeded in removing the negative rewards altogether. Even the metal plate survived her threats of pulling it up with the heel of her foot in protest.

We tweaked our study to resemble a human hearing test. When Donna answered correctly, we reduced the vibration level by 6 decibels incrementally. Then when she failed, we increased the level of vibration by 3 decibels, and if she answered correctly, we dropped it again by 6 decibels. In human hearing experiments, a threshold is established based on three of these reversals. With Donna, we used a set of fifty experiments, randomly giving a signal or no signal so that there were twenty-five signals and twenty-five times when no signal was presented. If she performed satisfactorily, we moved the vibration level down. Then we decided that two reversals would be satisfactory to determine the threshold of her ability to detect the vibrations.

There were, of course, a few technical challenges, particularly when we had to move down to lower and lower vibration levels, and Donna failed more and more often. She did not like to fail and got increasingly frustrated. Once, after failing several times in a row, she bent the target poles in half and threw them at us. With no more targets to throw, she turned her attention to the plate and attempted to suck it up with her trunk before being escorted away by a trainer bearing buckets of sliced apples.

After reburying our makeshift targets, we decided that we needed more serious ones. Colleen's brother, now the elephant manager for the zoo, welded steel target shapes onto steel railway tracks buried in the ground. This worked, and Donna was no longer tempted to remove the targets in her moments of frustration. Instead, she placed her trunk over the "start" target when she was ready to play the game.

Donna could always tell by our enthusiasm whether she passed each exam and could move on to the next level. She would bob her head, trunk swaying, drooling into Colleen's hand in anticipation of getting the next challenge correct and receiving an apple or banana slice, alfalfa cube, apple wafer, or bread treat. By the end of the trial, Colleen would be covered in elephant spit, and some of it often reached me, too. Donna's hairy, dripping trunk would fling forward with gleeful anticipation of the next treat, sometimes splattering the keyboard of the computer with half-chewed goo.

If Donna stood off-center, the unequal weight distribution on the plate sometimes caused the shaker to produce harmonics that were stronger than the test stimulus. Colleen would realign her by putting her arm up and asking Donna to "line up." "Line up, honey girl," she would ask sweetly. "Come on, Donna, line up." And Donna would happily comply by centering her feet evenly on the metal. What we found in time was that Donna did her best when placing her right foot over the shaker with most of her weight on that foot, a pattern that I had often seen in the wild. Fortunately, the posture did not compromise the integrity of the signal.

Donna often wanted to get on with the experiment much more quickly than we were able to reset it. She would get into a pattern of doing ten to twenty trials in a row, and then if I hit a technical glitch for a second or two, her rhythm would be thrown off. She would hit the middle target to signal to us that she was ready to go, and when we didn't respond, she would start to guess, hitting the "no" target since there was no signal delivered. When she didn't receive treats, she would get upset. It would take a little while to resettle her into the routine once the glitch was fixed.

We eventually got the routine down to a science and Donna was able to complete a session of fifty trials in about fifteen minutes. Then we reduced the decibel level and moved on. In the end, we were able to develop the fluency and momentum needed to complete an entire examination of six to seven vibration levels within a single session.

We realized over time that Donna did much better if she had shorter exam sessions with more breaks rather than the long string of fifty tests in a row. With shorter sessions, she seemed to be able to focus better and was a more willing participant. We were once able to conduct experiments for three days in a row, and she seemed more interested in success in the end than she was in the treats. This marked a new milestone in the experiments, as I hoped to be able to teach Donna to become more sensitive to lower and lower vibration levels.

As we witnessed Donna learning to focus on the vibrations as the signal got lower and lower, we could see the concentration in her face. She was so focused and intent on sensing the signal that when we finally established a working threshold measure, it almost seemed demoralizing for her to fail often during the exams by not being able to detect the vibrations. It got to the point where it was hard for Colleen and I to watch how dejected and confused she appeared by her repeated failure. We couldn't wait until that portion of the exam was over so that we could raise the vibration level and let her enjoy her success again.

During some of the more difficult trials, Donna seemed to transform suddenly from a jolly elephant, happily sucking her treats and making what elephant trainers call "rasberry" sounds which is sort of like a repetitive nasal clucking when she answered correctly, into a toothless, bearded hag, all hairy and cowardly, tentative trunk lips quivering as she reached toward the "yes" target, then kicking the heel of her foot into the plate when she got it wrong and didn't get a treat. She blossomed and withered, flourished and soured, depending on her success. It was an experiment designed purely to test vibration detection levels, yet it was also fascinating to witness these displays of elephant character and behavior. As we watched her swell and shrink, her suffering made us want to see her succeed all the more.

After one of her more difficult trials, we decided to give her a long break. We removed her from the lower bull yard so that she couldn't have access to the force plate, and once she got into the upper yard, she went into a play bout with such a sugar high that she sparred with Osh with particular enthusiasm, the bars between them giving him undue courage. He had been separated from the girls a few days earlier because Lisa was beating him up in the exhibit that was open to the public, so the keepers decided to give him a break. He had Donna racing up and down the enclosure, with Donna roaring, rumbling, rolling around on her back, tusking the ground, and then she came at him full force and was met head on with his young but growing brow and tusks. Donna would beat the daylights out of him given the chance, but he knew that he was safe in this situation. In a few years, the girls will figure out that the contest is over.

Before the end of Donna's play, some wild turkeys had strolled into the enclosure and Donna went trumpeting after them, sweeping her trunk at them, leaving them scurrying and gobbling away as fast as their legs could carry them. A little while later, she stood at the gate to the enclosure where the experiment was set up, letting us know that her break was over. She reminded us who was in charge at every turn. I was growing more and more fond of Donna the more time I spent with her.

Each day, there was a new lesson, a new insight into her intelligence.

What we had been able to determine thus far was that Donna was extremely sensitive to vibrations perceived through her feet and became more sensitive over time. But exactly how sensitive was still up in the air. In our most recent experiments, though, she had surpassed the level at which either Colleen or I could detect the vibrations when we stood on the plate. We were thrilled to step on the plate after the experiment ended to find that we couldn't feel a thing.

We were hoping to finish our experiments at the zoo in time to make an assessment of how far away an elephant could have been from the devastating earthquake that rocked Southeast Asia on December 26, 2004, and still be able to detect it. There were some remarkable reports from Sri Lanka and Thailand about elephants acting in a peculiar way around the time of the earthquake. There were also reports of elephants behaving strangely about an hour before the actual tsunami hit. Some were extremely agitated to the point of breaking their chains and running inland, away from the incoming wave.

Yet, at the same time, two elephants that happened to be radio-collared in Yala National Park, along Sri Lanka's southeast coast, showed no evidence of reacting to the event. This is most curious, as this was an area that was heavily affected by the tsunami, and as of yet, there is no explanation for why the anecdotes differed so dramatically from this scientific data.

If it was possible that some elephants did detect vibrations caused by the quake and the resulting wave, and indeed responded as the anecdotes suggested, the peculiar behaviors would have an easy explanation. Unfortunately, we weren't able to finish our experiments in time to make conclusive statements about the distance over which an elephant could detect these vibrations. But since there were reports of U.S. Geological Survey (USGS) researchers who felt the vibrations of the quake, it certainly seemed possible that elephants in some locations would have detected it as well.

In my preliminary investigations and from discussions with a colleague at the USGS, there were a number of factors affecting the seismic activity generated by the quake, as well as different possibilities for the types of noise the tsunami could have generated both in the air and in the ground. One important consideration was whether the wave was positive or negative. A negative wave would cause the sucking back of the ocean as was seen along the shores of Sri Lanka. This wave would create a very large amount of seismic activity simply because of the noise of rocks, coral, and other debris on the seafloor being rolled back into the ocean in a powerful current. Another factor was that most of the energy of the low-frequency vibrations was in the range of less than 1 hertz to a maximum of 10 hertz, so unless elephants were able to detect waves this low in frequency, they may have been detecting the harmonics caused by the disturbance or picking up on some other environmental cue.

People often ask whether elephants can provide humans with an early warning system for earthquakes. Judging from the anecdotes of the December 2004 tsunami, there may be other animals that would potentially provide much earlier and therefore more useful cues than elephants. It would also be impractical to house an elephant appropriately for such purposes. There is a scientist in Japan who believes that catfish can be used as earthquake detectors and that they provide enough advance notice that evacuation could be possible. Apparently this idea first appeared in Japanese folklore with tales of a giant namazu, or catfish, living in the mud under the earth. Although its destructive activities were kept in check by a god named Kashima, sometimes Kashima would be distracted and the catfish would be free to move about and cause earthquakes. Dr. Ikeya set out tanks of catfish in his lab and designed a computer program that monitored the movement of the catfish.

If the catfish moved outside a given set of parameters, the

computer would register this as a signal. Ordinarily, catfish tend to be stationary, and it didn't take much movement to generate a signal. This scientist believes that the buildup of pressure of granite prior to an earthquake generates electromagnetic pulses that catfish can detect. Catfish are equipped with electrical sensors, and since they communicate by means of electric pulse, they apparently increase their movement as the pulses become greater in rate due to an increase in rock compression.

Perhaps if we understood what cues animals are picking up, we could design our instrumentation accordingly. Paying closer attention to the electromagnetic environment appears to be a good place to start, and a team at Stanford started such a monitoring site at Jasper Ridge to collect better and more consistent data than had been collected in the past, which could be used to corroborate data collected at other sites. This was not the first time that electromagnetic sensors had been applied to earthquake monitoring, but in the past there were not enough sites monitoring quakes to get conclusive results, since unexplained spikes of energy had been recorded at odd times.

There are many possible ways for an elephant to detect an earthquake, the seismic environment being just one. Although the earthquake of December 26, 2004, was tragic, it was intriguing for me to hear the reports of elephants apparently responding to it. Despite the obvious caution with which scientists must view anecdotes, the incident provided a suggestion from an independent source—elephants might be capable of detecting seismic cues.

In the meantime, it was 10:30 P.M.—time to bring in the elephants at the Oakland Zoo. I walked up to the barn with Colleen as she pressed a few buttons to open the giant pneumatic doors big enough for a dinosaur, a red glow emanating from within the barn. She walked inside to check that the door between Osh's and Lisa's stalls was shut, because the night before the door had been left open by accident. Osh had let out a bloodcurdling roar when he found Lisa suddenly in his stall as she was being put away. She had marched straight from outside the barn through her stall and into his, immediately eating his food as quickly as she could before Colleen could run around and reconfigure the rest of the elephants while I fed M'Dunda bread slices to distract her.

But on this night, Osh was retrieved from the bull yard below the exhibit and he slipped quietly into his quarters and the door shut behind him. Then we walked past the barn and up the hill toward the exhibit, a three-quarter moon illuminating the landscape. The three girls were standing at the gate casting long shadows as they waited for Colleen to open it remotely from the side.

Colleen asked Donna to "line up" at the gate as she tossed her some bread slices from the gate-control area. As Donna was dominant, she needed to pass through the gate first so as not to cause any trouble with the other girls. Then Colleen called to Lisa, who was waiting in the wings some distance from Donna. As the gate opened, the elephants walked quickly down the slope from the exhibit to the barn, Donna first, then Lisa, and M'Dunda last. Donna had her own quarters, separate from Lisa and M'Dunda, on the opposite side of the barn from Osh.

Lisa crossed under the chain dividing her side of the enclosure from M'Dunda, an invisible barrier for her, but for M'Dunda, dominant to Lisa and a food stealer, it was as good as a brick wall. Their door stayed opened at the back, as they liked it a bit cooler than the others. We watched as Lisa reached through the bars to get at a 5-gallon water jug hanging just within reach that was stuffed with alfalfa pellets. Plants hung all around her in the dull red light, so you could make out her trunk poking through the yellow acacia flowers and leaves, delicately balancing the container against the wall so that she could tip it upside down, causing the pellets to fall out of the large hole at the top. She carefully scooped them into a pile and ate them enthusiastically. Little did she know that Donna had consumed several buckets' worth of these pellets, including an even more treasured treat, apple wafers. Earlier that day, Donna had gotten so full of apple wafers because of a new training regimen that she had even thrown a few at Colleen, letting Colleen know that she was in serious need of a time-out.

As Colleen checked to see that there was enough browse for the next feeding, I stood on the concrete and wondered how dulled to vibrations the elephants must be living right off the freeway. Perhaps Donna could be taught to be as sensitive to vibrations as an elephant in the wild, where human-generated seismic noise is not so much of an issue. Donna was learning to focus, and I began to think that if detecting vibrations contributed to her survival, she might perform better in these experiments. Perhaps a wild elephant would perform more like people with hearing impairments who have been shown to be more sensitive to vibrations than people with normal hearing and are also more accustomed to using vibrations as cues.

Wild elephants might need to focus on vibrations as part of their daily way of perceiving the world—just like a hearing-impaired individual. They might use them to communicate or make judgments about their environment. Such critical perceptions might include an elephant sensing that there is water in a particular location or another elephant at a certain distance; a hearing-impaired person might assess whether it is safe to cross the road.

Because we suspected that a wild elephant would be more sensitive, we planned to return to a reserve in northern Zimbabwe to work with trained elephants in a semicaptive environment. In this natural environment, there would be more reason to focus on vibrations as important signals. Having grown up in a relatively quiet environment, they would not be so habituated to human-generated seismic noises.

13

BACK TO MUSHARA
IN STYLE

One never notices what has been done;
one can only see what remains to be done

—MARIE CURIE

I STARED OUT FROM THE Mushara tower over the chilled landscape lit by a three quarter moon. The unusual cloud cover
obscured my view, smothering all but a few shafts of light that
stretched out to some far-off place in the east. When there were
gaps in the clouds, the moon cast a dim white glow over the
hushed land, turning on and off like a flashlight. It was as if a
surf were rolling in and out hypnotically over the silvery night
sand.

A lone bull stood at the waterhole in the quiet of the night,
keeping me company with his meditative silence. He, too, might
have been awaiting the arrival of a breeding herd. A long section
of his trunk lay flat on the ground, as if to eavesdrop on the con-
versations or footfalls of approaching friends. Or maybe he was
simply asleep. Just when I was convinced that he was sleeping,
he perked up, turned, and faced southwest with his ears out,
alternating between freezing and scanning the horizon. Then he
went back to resting. When he finally sauntered off, he sniffed at

our equipment hide, where we had buried the speaker and microphone under a pile of thorn bushes.

It was 2002, four years since I had been able to get back to the country, missing the war in the Caprivi in 1999 when some tourists were shot in their vehicles along the now fully paved Golden Highway. There had been gunshots from across the Zambezi, some innocent locals shot out of public paranoia, then the former pace of life resumed after a few brief skirmishes, impromptu road blocks, convoys, and other minor inconveniences.

In February 2002, just before Tim and I returned to Etosha, Jonas Savimbi, the rebel leader of the UNITA forces in Angola, was tricked and then shot in a skirmish, bringing an end to a bush war of almost thirty years. The elephants could finally breathe a sigh of relief. In 1992, Savimbi refused to recognize the results of the national elections and led his army back to the bush to continue fighting, sustaining themselves largely on bush meat for another ten years. In the meantime, the AIDS toll was rising inconceivably, and tuberculosis and other infectious diseases were reemerging in epidemic proportions with no relief in sight.

Because of the paper that my colleagues and I published showing that elephant vocalizations travel in the ground, my hypothesis about elephant seismic communication was gaining momentum and recognition within the scientific community. Media interest also helped to bolster support, but the research was slow and had to take a back seat to anything that could bring in a salary. Tim had started medical school, and we needed a steady income, so I cobbled together some research grants and postdoctoral appointments to keep me going.

I had the opportunity to immerse myself in the tools of molecular biology to understand vibration sensitivity and deafness at the level of gene control, providing a logical bridge with my interest in elephant seismic communication. I was excited by my mission and hoped eventually to apply my molecular skills to questions relating to elephants and the perception of sound, and perhaps even conservation.

It was around this time that I received several field grants to return to Mushara to explore elephant seismic vocalizations in the wild and whether elephants could detect these vibrations. So there I was, finally back at Mushara, but this time with a team of colleagues armed with a lot of seismic instrumentation and night-vision gear.

Because the research took place during the summer, Tim was able to join us, excited to take charge of some of the technology. He had learned how to use a differential global positioning system (DGPS), which we needed to get exact distance measurements between our sensors. I coordinated the geophysics instruments and the experimental design while a graduate student in geophysics ran the clunky old seismic UNIX-based instruments that we had borrowed.

We no longer had to rely on the Mushara bunker for protection. We now had a spectacular view from a tower over 20 feet above the waterhole and 400 feet away, which meant that we could monitor the comings and goings of the elephants much more effectively. The tower platform was an 8-foot-square space, with shade cloth walls about a meter high. It was a tight fit for all our equipment and electronics, as well as chairs for ourselves, but we made it work.

We surrounded the perimeter of our camp under the tower with what the locals call boma cloth, or game capture netting, an opaque shade cloth that animals view as a solid object. This allowed us to move freely and safely around the camp and not disturb the animal traffic to and from the waterhole. The barrier also provided some protection from the wind and was an effective way of keeping lions out of the camp and jackals out of the kitchen. Finally, we were far enough away from the elephants that our camp noises had less effect on them than from the bunker. It was now possible to have an entire team at the site. Even so, I was fairly conservative about the amount of noise that we made that season, not knowing how much the elephants would tolerate.

I chose Mushara as the site for continuing my seismic studies because the alarm calls we planned to play back through the ground were the ones that I had recorded there. The elephants recognized and responded to these recordings when I played them through the air, so I wanted to see how they might affect the elephants if I played them back through the ground via specially adapted transmitters, or "shakers," buried near the waterhole, similar to the one that Donna stood on for her vibration sensitivity exams.

We spent the first three weeks setting up the site and col-lecting data, after laying down and burying our six instruments over a course of half a mile. Each contained three channels, which allowed us to space the geophones at 50-yard intervals from the waterhole into the bush to record elephant rumbles as they traveled through the ground. We recorded rumbles using microphones to measure how far and how fast the elephant vocalizations traveled in the air, and then a similar recording device, a geophone, to measure vibrations in the ground.

We hoped to model how far the rumbles could travel, given the strength of the rumbles at the most distant geophone, and the fall-off rate between sensors. Our previous measurements were based on a distance of only 100 feet from the elephant (such were the limitations of our equipment and the noisy ground environment in our Texas experiments), and we hoped to get a better idea of a real fall-off rate of the signal over distance. We were also hoping that we might better ap-proximate an ideal propagation distance in the wild because the park was made predominantly of calcrete, making a very homogeneous medium for vibrations to travel with little to no obstructions. In fact, it was probably the most ideal set-ting we could have hoped for, similar to the sound fixing and ranging (SOFAR) channel in the ocean or an inversion situa-tion in the air, where sounds are forced to travel in two dimensions rather than three, providing a clear sounding board for up to 6 miles.

Starting in the late afternoon when elephant family groups visited the waterhole, I took notes on when a rumble occurred and in what context, especially if it was a loud rumble and if it occurred at the head of the trough. We had painstakingly measured out each geophone from the head of the trough with the DGPS, providing the centimeter accuracy needed in our distance calculations. It was difficult to record anything in the daytime because the family groups hardly came in before 4:00, and although bulls visited throughout the day, they did not vocalize much. The wind also made it challenging to get a good acoustic recording during the day to compare to the seismic recording.

I tried to focus my data collection when there was little activity from other species at the waterhole in order to minimize the amount of background noise in the ground during an elephant vocalization. Normally, a "let's go" rumble and the responses that followed were not as loud as vocalizations made out of aggression toward another herd, a young bull, or a rhino. But if these dialogues were initiated while the matriarch was at the head of the trough, I noted them because they were usually made when the herd was calm and the conditions were quiet.

I put stars next to the rumbles that were particularly loud, like when a matriarch voiced her objection to a rhino. These elephant–rhino interactions happened often at the waterhole because elephants don't like to share their drinking time with other species, and rhinos, territorial animals that they are, tend to hold their ground and insist on sharing rather than waiting for the elephants to finish their waterhole visit before approaching. Or if a young bull was causing chaos within the family, he'd start to rumble, a poke here, a whack with the trunk there, resulting in a wailing youth roaring off, perhaps letting off some steam by picking a fight with a rhino mother drinking at the edge of the pan with her calf.

In the beginning, middle, and end of the recording, we positioned microphones to record the rumbles. We made protectors for the microphones using PVC plumbing because the fuzzy black windsock covers proved tempting to lions as a toy and because they would be less likely to get crushed by an elephant. Unfortunately, our attempts to camouflage the expensive microphones were foiled when a pride showed up one morning at dawn and the mother took the irresistible toy for her cubs.

Tim and I had been sleeping in the tower and woke up to see these beautiful animals in the predawn light. We watched confidently at first as the lioness pawed at the thorn bushes that protected the microphone placed near the waterhole. She lay flat on the ground and reached as far as she could underneath the thorns and around the rocks, trying to get at the tempting fuzzy object. Suddenly our confidence turned to horror as she extracted the mic from the thorns. With a swift swipe she pulled it out, and in no time the cubs were batting it around the waterhole.

Tim raced down the ladder and jumped into the truck. He hurtled toward the pride and was met by the male and female standing firm in front of the hood, refusing passage. Tim honked the horn to no avail until the female grabbed the microphone and took off into the bush. In hot pursuit, cursing and swearing the whole way as he hit a ditch here, a log there, Tim followed. He eventually cornered the lioness, which sat for a long time holding the catch in her mouth. Eventually she dropped it and ran off. We were relieved that our vital piece of equipment had been rescued.

With the first experiments complete, we packed up the seismic instrumentation, burned twenty DVDs' worth of data, and sent the first crew off with instruments and disks in hand. We had a week before the second crew would arrive and we could begin to conduct the seismic detection experiments. In this downtime, I brought up the possibility of measuring elephant footfalls as well as those of other large mammals in order to develop a seismic censusing technique. Elephant footfalls had a distinctive pattern that was easily measurable with our geophones, and so did other large mammals such as the giraffe, kudu, gemsbok, rhino, and lion. It wasn't unreasonable to think that footfalls could be used as a tool to monitor the movements of these different species, collecting potentially valuable data on who visits the waterhole and when.

This new censusing technique could be placed at remote waterholes, and data on the numbers and timing of visits could be gathered. The data could even be sent to ranger stations remotely via satellite. It was even conceivable to design an instrument that would detect elephant footfalls at the perimeter of a farmer's field and set off an alarm to alert farmers to the presence of elephants.

Excited by this goal, we decided to run a geophone off the tower at a 50-yard distance and collect data from footfalls for the rest of the field season. We amused ourselves with the different ways that animals walk, the elephant being the most curious. The back leg moves first, cushioned heel to toe, then the front leg on the same side moves forward, allowing space for the back foot to be placed in the same spot as the front footprint, then the pattern is repeated on the opposite side. Moving both feet on the same side and then switching to the opposite side gives the impression that elephants are sliding forward on a Nordic ski track. Hence these enormous animals can look like they are floating.

Our blissful week of rest went by quickly. The new research assistants arrived. The most difficult challenge of the next experimental setup was making sure there was no evidence of the seismic playback in the air while the signal was transmitted through the shakers so that we could be sure all responses to the playbacks were from detection of the signal through the ground via the elephants' feet and not through the air via their ears.

It was 11:10 p.m. and my shift ended at 2:30 a.m. The cold was just enough to keep me on edge if I kept moving every few minutes to scan the horizon for elephants. Without moving around, I risked getting a little too comfortable in my blanket, drifting off to sleep, and missing a herd crossing the clearing to arrive at the waterhole undetected. Ideally, it was best to sight a herd as they neared the edge of the clearing in order to wake my colleagues and turn on all of the instruments to prepare to conduct an experiment, as well as to videotape their approach, which was the best way to count them and provide enough documentation to later establish the size differences between individuals. By the time a herd reached the waterhole, they were generally too congregated to get a good idea of age distribution.

Our nightly routine for the playback studies was simple: we documented the time of arrival of the herd at the waterhole, then waited two minutes for the elephants to settle down before starting a five-minute control period to monitor their behavior prior to playing back the alarm calls that I had recorded years earlier.

We played back the calls seismically once every minute over the course of three minutes, and then we watched responses over the next five minutes: vigilant behaviors such as freezing, scanning, smelling, foot lifting, and vocalizing. We were also keeping track of the time the herds would spend at the waterhole during a visit that included a playback experiment versus visits when no playbacks were used.

I finally heard a soft rumble from behind me on the southwest path. I turned around and saw a small herd standing at the edge of the clearing. I turned on the walkie-talkie and alerted Jason Wood, my postdoc at the time, who then woke the others to gear up and get to their posts as quickly as possible. I went through the list of night-watch duties: turn on the video camera and night vision; turn on my monitoring computer, then Jason's playback computer; and start counting elephants if they arrived in the clearing before the others had reached their positions.

I heard the others wrestling with their warm gear from within the roof tents, as well as the soft swishing of leather from the herd briskly lumbering past our camp. While videotaping their approach, I dictated the herd composition into the video microphone.

Once everyone was in position and the elephants had settled down, we were ready to begin an experiment. "Start," I whispered into the microphone of the video recorder, as the time on the videotape and the real time were recorded. Temperature and wind readings were taken during the five-minute control period. Then Jason hit the play button on his computer, the source of the seismic alarm calls. I took notes on behaviors, switching between low-light binoculars and a panel on the computer screen rigged with a view of the night-vision video recorder.

Not knowing exactly what to expect from the elephants in response to the seismic component of the call, we designated five behavioral characteristics as signs of heightened vigilance: freezing; smelling the air with trunks; leg lifting; leaning; and orientating toward the source of the signal. We used three parameters to examine herd spacing: greater than one body length apart; one body length; and less than one body length.

It was amazing to watch the breeding herds in action during our trials. They didn't run away as they had done with acoustic trials. Yet I watched the night-vision illuminated green monitor on my computer screen in awe, as whole herds would freeze in unison. If they were drinking, they would all stop and be silent as the trunks released water and hung in front of them. If they were walking, they froze midstride, extremely focused, shifting their weight and leaning forward.

Then slowly, almost imperceptibly, the elephants turned themselves toward the shakers, and then finally each family bunched up into their individual groups, a defensive posture, the little ones tucked safely in the center of the cluster. Minutes before, they had been spread across the length of the trough and halfway around the pan. At that moment, they all pulled together, as if they were all connected, like one giant marionette controlled by strings from above. I was astounded and thrilled to see this coordinated behavior. The entire herd was reacting, just as I had noted years ago when I saw whole groups freezing prior to the arrival of another herd.

My heart pounded. All these years of planning, hoping, and dreaming of this moment. We were finally showing that my original hunch so long ago was true. Elephants were detecting and responding to our seismic cues. I tried not to get too excited, as I knew that we would still have to prove the finding to be true statistically. Sometimes patterns are seen in nature that don't turn out to be significant, and there were certainly some experiments that were more dramatic than others, so I held my bias at bay and moved forward, repeating as many experiments as possible in the hope of obtaining significant results.

Besides turning toward the shakers, the herds tended to line up perpendicularly to the shakers. Later we realized that this orientation might indicate how the elephants perceive seismic cues, by hearing or by touch. If an elephant were to stand perpendicular to the shakers, the posture would allow a greater distance between the two ears, so bones in the ears might be the pathway for detection. Alternatively, if they were to orient parallel to the shakers, there would be a greater distance between the front and back feet, which would indicate that they might be feeling the vibrations. Certain conditions could favor one pathway over another, but in this case, the elephants consistently oriented in a way that would facilitate a bone-conducted pathway.

In watching the responses, it was as if they thought trouble was in the distance, as though the herd couldn't identify the source or the exact nature of the trouble. It would make sense that without the airborne component to the call, the elephants would perceive such a signal as being a distant event.

After crunching numbers, we found that elephants spent significantly less time at the waterhole during seismic trials. They bunched together and consistently oriented in the direction of the source but perpendicular to it. They also increased their vigilant behavior, although that wasn't as significant.

In the end, we were able to demonstrate that elephants can indeed detect vibrations. However, it took several years to get our work published because the study was so interdisciplinary and thus difficult to find appropriate reviewers. Usually, once a precedent has been set, a scientist refers to earlier studies that established such and such or refers to methods designed by so and so, but we didn't have that luxury and had to start from scratch.

As for how far elephant vocalizations traveled through the ground, all that was evident in our data files was the air-coupled component of the signal. I found out the hard way that you can never ask too many questions of too many people in a particular field and that there are never too many ways to ask the same question of each expert. Eventually, a colleague at the U.S. Geological Survey assessed the situation and informed me that we had spaced our geophones too far apart and this explained why we were not able to see any of the groundborne signals in our data set.

"It's a simple anti-aliasing problem," he shrugged. Although he provided a wonderfully eloquent definition of the problem and I had certainly come across this term in my acoustics research, I still didn't really understand how it applied to the seismic environment. I had to look up what anti-aliasing was in order to learn its meaning in this context. Apparently, the terminology is used in other fields, and I found the computer graphics explanation of aliasing most useful visually.

If you were to imagine a black-and-white checked pattern going off into the distance, aliasing means that the checks in the foreground and the checks in the background are given equal weight in a graphic image, causing a jagged or block pattern when representing a high-resolution image at lower resolution.

These jagged lines are caused by different continuous signals becoming indistinguishable, or aliases of each other. And in our context, all of the seismic signals present in the ground had become indistinguishable in our measurements even though they were traveling at a different rate. So because our geophones were placed apart at 50-yard distances rather than at a distance of 3 or up to 10 yards, we were sampling at too low a resolution to distinguish the low-frequency surface waves in our data from the other higher-frequency waves.

We later solved this problem by repeating this experiment at a private facility in California using fifty-seven geophones spaced at 3 yards over a distance of 180 yards. In this study, we demonstrated that elephant vocalizations indeed propagate along the surface of the ground, like ripples along the surface of water. We used a much higher-resolution technique than we had used previously in Texas, a considerable advancement over our original paper on the subject.

HAVING SEEN OUT the cycle of the moon, it was time to pack up Mushara for another season and head back to the lab to pursue questions relating to anatomical features that might facilitate the detection of vibrations. I soothed myself with the promise to return.

14

❦

KEEPING AN EAR
TO THE GROUND

*A certain raja presented an elephant to a group of blind
men. "Here is an elephant," and to one man he presented the head, to another its ears, to another a tusk, to
another the trunk, the foot, back, tail, and tuft of the
tail, saying to each one that that was the elephant.
When asked what the elephant was, the man presented
with the head answered, "Sire, an elephant is like a
pot." He who had observed the ear replied, "An elephant is like a winnowing basket." Presented with a
tusk, another said it was a ploughshare. He who knew
only the trunk said it was a plough; another said the
body was a granary; the foot, a pillar; the back, a mor-
tar; the tail, a pestle; the tuft of the tail, a brush. Then
they began to quarrel, shouting, "Yes it is!" "No, it is
not!" "An elephant is not that!" "Yes, it's like that!"
and so on, till they came to blows over the matter.*

*. . . For, quarreling, each to his view they cling.
Such folk see only one side of a thing.*

*—*UDANA *68–69: Parable of the Blind Men and the Elephant*

ELEPHANTS HAVE TWO POSSIBLE pathways to detect vibrations: either through bone-conducted "hearing" or through a sensory pathway not connected to the ear. And sometimes, both. When vibrations travel through bone, they move through the feet, then the legs, shoulders, and finally into the middle ear cavity.

Elephants have three middle ear bones that are enlarged; one, the malleus, is especially enlarged. The additional mass of this bone facilitates the independent oscillations of the middle ear bones relative to the skull due to inertia. The process is equivalent to movements caused by sounds entering the ear canal, hitting the eardrum, and stimulating these bones directly. Bone conduction, in effect, skips the eardrum step. (A simple example of this effect can be demonstrated by wearing earplugs. Your own voice will still sound fairly loud because you are hearing it through bone conduction.)

In addition to the enlarged ear bones, the dense fat in the footpad may also facilitate a bone-conducted pathway. And a third feature is the use of a skeletal muscle in the ear—what Byron, our field geophysicist, had termed "ear lips."

But conduction is not the only way that animals detect vibrations. There is another pathway through specialized sensory cells throughout the body rather than via the ear. In primates, there are three different vibration-detecting cells found in the lips, hands, feet, and muscle lining the intestine. Each is capable of a range of detection. These same receptors can be found on the tip of the Asian elephant's trunk, and are thought to be as sensitive as the lips on a primate.

In fact, many small and large mammals are equipped to detect seismic cues this way. One notable example is the star-nosed mole, whose snout is surrounded by fleshy appendages covered with thousands of organs that act like a tactile eye. But specialized vibration-sensitive cells have also been found in the paws and knees of cats, in the knees of kangaroos, and in the beak skin and knees of birds to help them find prey and avoid predators.

To investigate just where specialized cells are located in an elephant's foot, my colleague in Berlin, Thomas Hildebrandt, kept an eye out for elephants that died in captivity. Eventually, he was able to secure two specimens from the Berlin Zoo: an Asian elephant and an African elephant. Although his facility could accommodate very large animals, manipulating a whole frozen elephant was quite a feat. The weight of an adult head alone required handling with a grappling hook before we could isolate the areas of interest—mainly the ear and the sphincter-like muscle at the opening of the canal.

We found vibration-sensitive cells in the feet similar to those found in the trunk. These cells look like onions, with many layers surrounding a nerve. The layers shift in response to a vibration, sending a nerve impulse to the brain. One of my students back at Stanford mapped these cells in sections of the bottom of the foot and found that the majority of them occur at the edges, mostly at the toe and heel, possibly explaining why the elephant rolls forward or back on its front foot when it detects seismic waves.

We investigated the structure of the cells because we were having trouble with what a colleague had called the "lemon effect"; that is, it was difficult from thin sections of tissue to piece together the size and orientation of the cells. If you were to take a dozen lemons, place them in random positions in a confined area, and then slice through them, as you would when preparing a slide of elephant foot tissue, you would come out with very different shapes and sizes of slices, depending on the position of the lemon within the space. Unless you were then to piece together all the slices on top of one another, you wouldn't be able to determine the shape and size of the cells.

We were hoping to solve this problem by reconstructing the three-dimensional volume of the cells. We wanted to see if elephants had unusually shaped cells that might allow them to assess the force of vibrations.

AT THE SAME TIME as these anatomical studies, I geared up my research team to return to Namibia to continue with our seismic playback studies, this time to determine whether elephants could discriminate between seismic cues. During our supply run to Windhoek, I stopped in to see Simon, and he told me that one of the first resource monitors I had hired in the Caprivi had died of AIDS that year. I wasn't prepared for this news. I should have expected that it was only a matter of time before AIDS hit my inner circle of colleagues, but I had kept my wheels of denial well oiled on this topic. Now there was no way around it. AIDS had reached epidemic proportions, with a rate of 40 percent estimated for the Caprivi and possibly higher. It took several days for this news to sink in.

I wasn't able to get to visit with Janet, so I asked Simon to convey my condolences. I knew that Janet would be very hurt if she had heard that I was in her country and didn't try to see her, but it was getting too difficult to get to the Caprivi after the fieldwork in Etosha was completed. I had hoped that there would be a chance the following year to see her.

There were several improvements to the camp that year, as well as the tower and Mushara cafe, the affectionate term for our starlit bush mess hall that was fast developing a five-star reputation with at least ten different recipes for how to prepare a butternut squash. We built a second floor on the tower to provide the blind observers with a comfortable vantage point for data collection. The second improvement was to mount four roof tents on 7-foot-high poles within the camp to provide extra safety in addition to affording an extraordinary view.

Our Bushman colleague, Johannes, from Etosha Ecological Institute, was helping with the construction and laughed at our roof tents perched on poles. He said that they reminded him of the nickname for his tribe of Bushmen called Hicum. Apparently, they were called "tree sleepers" because his ancestors made smoky fires at night and slept in the trees during the wet season to avoid malaria-carrying mosquitoes.

These roof tents freed up the tower to be a fully functioning research lab. Despite various camp melodramas over such urgent matters as not wanting to share a tent, snoring, and the mysterious case of one individual who repeatedly lost personal items, all the research assistants were eager participants, even in the frigid cold at one o'clock in the morning.

What I wanted to determine that year was whether elephants could distinguish between seismic call types, and between familiar and unfamiliar callers. I could then begin to test whether elephants may in fact communicate complex messages through the ground. And, of course, I was interested in determining whether some components of alarm calls were universal, which would mean that elephants would be less likely to get used to signals used as crop-raiding deterrents.

I planned to play back familiar and unfamiliar alarm calls to see if family groups could distinguish between an alarm call made from an individual they recognized and one that was made from an elephant in a different country. Then we would compare the responses to human-made calls such as synthetic warbles (used in human hearing exams and Donna's studies), as well as trials in which there were no signals presented at all. I also planned to play back seismic estrus calls to determine how much bulls would pay attention to ground-born signals.

One last goal for the field season was to perfect our censusing technique measuring footfalls. We began by putting out geophones in a triangular pattern, each spaced 10 meters apart, which we suspected would be the ideal design for recording the footfalls of elephants and other large mammals. This time, we used a much more powerful seismic instrument called a Geode. We upgraded our shakers from those powerful enough to shake your car seat to shakers called "buttkickers," designed to be mounted to your house to enhance your home theater experience. Since these shakers resonate at 5 hertz rather than 60 hertz, I was hoping to have a better chance of transmitting the lower frequencies with more energy than in the past. Having a single seismic instrument to contend with was much less work. And we were able to operate it from the tower.

I hired some of the ranger staff to help us dig the trench for the cabling before the rest of the research crew arrived. And once all the cabling and geophones were placed, Jason and I did some calibrations, one of us taking measurements while the other jumped between sensors, then compared our results. I was inspired by a study reporting differences in footfalls not only by sex but also by "stooping" rather than walking upright. Subtle differences between gaits and sex were measurable, giving me confidence that we might ultimately be able to detect differences between individual animals.

When the rest of the team arrived, we settled into preparing the camp and the research station. The first thing I wanted to do was to get the bull studies up and running. Since the bulls spend such a long time at the waterhole during the day, it was an ideal opportunity to conduct playback studies on them. I knew that they did not respond to the alarm call playbacks, so I had to come up with something more enticing. One of my colleagues, Joyce Poole, had recently completed a study on musth bulls and estrus calls, in which she got bulls to respond to playbacks of estrus calls. She agreed to lend me two of her estrus sequences, one from an elephant named Shirley and one from Erin.

I had invited Colleen Kinzley along to help with the seismic detection studies, and with all of the downtime during the day waiting for breeding herds to arrive in the late afternoon, I had given Colleen the task of starting a bull identification book. We needed the ID book to be able to recognize distinguishing features, the age of the bulls with which we were experimenting, and what their state of musth was, if any.

Once we had a working knowledge of the bulls, we were ready to test our estrus call playbacks. Joyce Poole had found that nonmusth bulls would not have an interest in the estrus calls and that older musth bulls would ignore the calls if younger, more aggressive musth bulls were present, not wanting to contest the more vigorous competitor. We were careful to make sure the appropriate scenarios were in place prior to playing back estrus calls.

At one point, an older musth bull we'd named Gray was at the waterhole along with the adolescent bull named Osh. We decided to conduct an acoustic playback to see if we could generate any interest from Gray without other adult bulls present. Osh was extremely interested in these calls and approached the speaker in response to each playback, while Gray did nothing but remain on the other side of the trough, occasionally engaging in a trunk twist over the front of his face, typical musth behavior.

We played the calls back again and again, getting Osh more and more excited as he ran back and forth in front of the speaker, a confused yet eager young bull. We couldn't help but laugh at his confusion. After some time, it was clear that Gray had no interest in our presentations, but we did seem to be attracting new young bulls. When a breeding herd arrived, we decided to abort our mission for the moment to focus on the experiments with family groups.

We randomized our playback selections for the family group experiments and kept track of which herd received which seismic presentation. Our system ran just as it had the previous season: two naive observers on the lower level, one observing behaviors and the second observing herd spacing and individual orientation. The observers, Dave and Colleen, had their jobs down to a science, a stopwatch beeping every fifteen seconds to alert them to take the next scan sample. Colleen and I were in radio contact in case there were any questions about whether to stop a trial if another herd arrived in the middle of an experiment and to alert me when a fifteen-minute data collection period had ended.

On the top floor of the tower was the technology center. Jason ran the Geode, collecting data from the microphones. As usual, the videographer not only recorded the trial but documented herd size and composition, wind speed, temperature, and the time of arrival and departure of the herds. Tim was on hand to troubleshoot the various inevitable technology failures.

I got to watch all the action on my computer through the night-vision feed from the video monitor. I also took notes on behavior, any particular herd characteristics, and any potential environmental influences on the trials, such as how many rhino may have been present or if there were lions in the vicinity. Sometimes, what might have seemed like the perfect scenario for a single-herd experiment would turn into all-out chaos in the middle of the trial as two or three more herds would arrive to disrupt things. On these occasions, Colleen would call up to me on the radio and ask if I wanted them to continue collecting data. She and Dave were always greatly relieved that we decided to call off the experiment, because it was impossible for them to tell who was in the original herd amid the chaos of fifty additional individuals mingling about.

The most thrilling pandemonium occurred during the new moon. It was pitch black and Dave was in the tower on watch. We were all below sipping Amarula, the local equivalent of Bailey's, made from the marula fruit, before retiring to the warmth of our sleeping bags. A large herd came screaming and trumpeting in, and Dave reported through the night vision that he thought there was some kind of a fight going on. We all ran up to the upper level to see what was happening.

In front of us, scattered about in the clearing and clumped up in the western quadrant, a mating pandemonium was taking place. A cow was in estrus and was being chased by two males that were battling it out for her attention, but it was hard to tell if the bulls were in musth. They were at least a similar size—that much we could tell. It was so dark that even our night vision was not that helpful. We tried shining an infrared spotlight to enhance the night-vision intensifier, which gave us a little more resolution.

The bulls chased the cow around the clearing. Although cows apparently prefer to mate with older bulls, in this case it seemed like the cow was going to receive the winner of the duel. But without being able to identify the bulls or the cow through the green fuzz, we were not going to be able to piece together the real story.

For the moment, it was just very exciting to hear the jubilant screams and roars of the herd as a mating event occurred about 100 yards away. The winning bull stood upright, front feet perched on her back, while he coupled with his mate. A crowd had amassed around them, rumbling uproariously for the fifteen seconds or so that the copulation lasted before the bull got down and the chase continued.

We guessed that the next few days were going to be like this for the cow until she finished ovulating. To complicate the situation, the cow emits a "postcopulatory rumble" after copulation, which is thought to invite a challenging bull that might be bigger and therefore preferred. Unfortunately, the whole activity moved into the bush and off into the distance, so we'd never know whether an even bigger bull arrived on the scene.

After this excitement, we called it a night, as Dave's shift was almost over and it had dropped to 0°C. Besides, it was too dark to conduct any playbacks.

The next day, I logged the mating pandemonium into my journal. The journal-style notes that I took during trials and of particular events over the course of the month often helped me to re-create events and saved time having to go back through the original videos. But when all else failed, returning to the videos was essential to re-creating the scene. The videos were also crucial for any analysis that couldn't be completed during the experiment.

Over the course of these experiments, it seemed clear that elephants weren't particularly interested in our artificial warble-tone playbacks. And yet it was difficult to tell at first whether there was a clear reaction to the unfamiliar alarm calls. It seemed that families with more skittish matriarchs responded, but those with more confident leaders ignored them. This made sense, since McComb had shown that less savvy matriarchs would respond nervously to unfamiliar contact calls while the more savvy ones would not be worried about the prospect of contact with an unfamiliar family group. I made notes on this observation throughout the month to see if the pattern became more evident.

The familiar alarm was eliciting the response that we had hoped for, although as we were getting to know the family groups better, some were responding more intensely than others. This was where our individual herd identifications were becoming more and more important. In addition, we didn't want to repeat our experiments with the same animals, trying to avoid the problem of artificially inflating a particular outcome. It was important to figure out who was really going to care about these calls, and perhaps at some point in the future, we might hope to ask why one herd might care more than another.

Over the course of the month, we were also able to collect great data for our seismic censusing technique. We were getting elephant herds of different sizes, as well as many individual bulls, and good data on other species such as the kudu, lion, gemsbok, giraffe, and even humans. It was interesting to see how similar the human footfalls were to the elephant's. As I learned later, if you were to take off the spongy platform shoes that elephants wear, the elephant and human foot are remarkably similar on a CT scan.

OVER THE COURSE OF THE MONTH, it was becoming clear that my suspicions about this bull population were correct. My time spent in the bunker taught me that something quite remarkable was going on in this secret society, and I wanted to get to the bottom of it now that I had the luxury of more help and daylight hours.

Elephant bulls were usually thought to form loose associations or to be solitary. This description did not seem to fit the patterns of bull interactions I was observing at Mushara, since it appeared that many bulls had very close bonds. Nor did some of these bulls seem aggressive or intolerant toward each other when in musth. This was not how other researchers had described their observations of bulls in other regions of Africa, where musth bulls were reportedly intolerant of all other bulls around them, rising in rank to a dominant position during this heightened hormonal state. This was just one more mystery I hoped to shed light on over the course of my studies in Etosha. What was really going on in this apparent "gentlemen's club"?

I had also noticed that bulls seemed to coordinate their visits to the waterhole, often a mile or so behind a breeding herd or other bulls. When one showed up, there were bound to be more arriving soon. Most bulls, in fact, seemed to travel in discrete bachelor herds, whereas a few preferred traveling on their own, sometimes meeting up with herds during waterhole visits. And the bulls actually appeared to spend more time freezing and leaning forward on their front feet than the females did, outside periods of danger when the cows seemed to be more tuned in. They also used their trunks more often during these periods, laying up to a meter-long section on the ground to "listen" to distant signals through the ground.

While I was trying to make sense of these behaviors, I realized that the younger bulls sought the company of the older bulls, and whether the older bulls chose this or simply tolerated it was still unclear. It was also not clear whether the older bulls truly allowed the younger bulls to travel with them or whether the younger bulls simply timed their arrival at the waterhole to be around the older bulls. Some of the younger bulls would leave the waterhole with the older ones and continue on the same path for some time, but I wondered what eventually became of the camaraderie when it was time to share a favored dinner spot.

Along with the identifications, Colleen also noted any interesting behavior, and it made us realize that bulls in musth never challenged the dominant bull at the waterhole. Musth didn't have the expected effect of allowing less dominant bulls an opportunity to rise temporarily in rank to the dominant position.

Colleen was able to identify fifty-nine bulls that season. She fell in love with her task. Any time there was an incoming bull, she would rush to the upper level, grab the ID book, and try to figure out who was visiting us.

On the last night of the season, Colleen and I decided to sleep in the bunker—me for old-time's sake and because I knew that Colleen would find the opportunity to get up close and personal with the bulls a memorable experience. After we had settled in, we had a visit from the don and his gentlemen's club just as the sun was setting. The younger members arrived first, then stepped back to make way for the don at the head of the trough. Each young bull returned, placing his trunk in the don's mouth, one after the other, upward probing trunks silhouetted against the pink sky. Later when we gave up our vigil, I was soothed to sleep by the slow breathing and leisurely drinking of several bulls that came for a long visit in the still of the late night.

15

WYNONA'S LAST STAND

*When you have got an elephant by the hind legs and he
is trying to run away, it's best to let him run.*

—ABRAHAM LINCOLN

THE DARKNESS CREPT UP ON ME while I was working on the
computer in my roof tent. A cow from an arriving herd bellowed, perhaps objecting to the light from my computer screen,
which I had so rudely neglected to cover with a red filter. I
hadn't wanted to get up, still recovering from a twenty-four-
hour bug that had been making its way through camp (four
workers down and three to go).

I turned off the computer, then turned on the night-vision
monocular to get a count; the first herd often arrives under the
cover of darkness. There were twenty-two with a tiny baby. This
was not Wynona's herd, nor Margaret Thatcher's, nor even Slit
Ear's. And no bull accompanied them. I took note of the time,
6:46 P.M., June 24, 2005. It was getting close to the end of
another field season.

Each year when I return to the site, I test-play the alarm call
to see if it still has the same effectiveness before playing back the
seismic version of it. The results from the previous year were
even more conclusive than I expected. Not only were family
groups able to sort out the meaningless signals, but they also

distinguished between familiar and unfamiliar alarm calls transmitted through the ground. I was bowled over by the results: although I had hoped to be able to show a much stronger reaction to the familiar alarm call, I didn't expect a total lack of response to the unfamiliar alarm.

This year, we were starting to see differences in the responses between herds to the familiar alarm. I suspected that either some herds had gotten used to it or that responses depended on whether a particular matriarch was recognized as giving the call, or simply that each matriarch had a different experience with lions and perhaps a different level of wariness in general and hence would respond according to her own character and history.

Now that I knew the character of some of the matriarchs and the relationships between the herds, I decided to conduct more acoustic playbacks than I had planned to figure out what was happening. I wanted to confirm that the same result would hold true if all three call types were played through the air. We were still using the same alarm call that I had recorded in 1994, as well as a similar call made by a breeding herd in Kenya. The third call was a warble tone, which we used as a control.

It didn't take long for the research to settle into a nice routine. We had worked out the kinks of our data collection technique for the bull study, and we were slowly building a good photo ID book for the bulls and breeding herds, as well as building our playback samples. When we weren't conducting acoustic experiments, I simply kept track of the time that the herds spent at the waterhole, how many individuals there were, and herd composition (full-grown, three-quarters, half, one-quarter, and less than a year, when the babies still fit under their mother's belly).

We were also getting better identifications of physical features within the herds. This was important because it was sometimes difficult to distinguish herds based on their size and composition, particularly during a new moon; our night-vision equipment was helpful but did not perform that well in near-total darkness. During these "time at waterhole" controls, we spent our time scouring the herd for any distinctive physical features that could help us identify them.

THE HERD AT THE WATERHOLE took off with a bellow and a deep rumble at 7:01 P.M., annoyed to be pushed off early by the impatient thirst of newcomers. I looked and counted once again. It was Slit Ear approaching on the heels of the leaving herd. Slit Ear's herd appeared to have the most dramatic response to the alarm call playbacks. Perhaps she was related to the matriarch that had made this call eleven years earlier. We really needed more samples of her herd's responses, but I was still sick and was not up to experimenting that evening.

Wynona's herd, in contrast, did not respond so dramatically. Perhaps eleven years was enough distance for her to forget that matriarch, or perhaps she was not affiliated with that herd to begin with and therefore the call would not have much meaning to her. Or Wynona just plain had no fear. I had grown fond of Wynona; she showed uncommon bravery, and yet, unlike Margaret Thatcher, she was adaptable and accommodated other herds at the waterhole.

Wynona was a young matriarch with a rough W cut into her left ear, and she was missing a left tusk. The year before, she was just a regular visitor, but now we had identified her and even obtained dung samples from her and two other herd members. It was in following Wynona's antics that I began to wonder if the matriarchal society of elephants might actually involve some kind of group coalition rather than a benevolent dictatorship. Other scientists thought that the dominant female within a family group was the oldest female and not necessarily the daughter of the previous matriarch. But Wynona was at least the fourth oldest cow in her herd, yet she was the one who clearly made the decision to approach the waterhole, leading the group in when it was safe. When she decided to leave, however, she would stay behind and bring up the rear.

Unfortunately, we didn't know the history of this herd, which might have helped us understand Wynona's rise to power. None of the other older cows appeared ill, so Wynona may have stepped in as the daughter of a matriarch that had died and passed along her status. Or perhaps she was especially vigilant, and the other cows saw her leadership as advantageous to the group.

Our second playback to Wynona's herd occurred on an evening just after two lion two-year-olds had settled in at the waterhole, a brother and a sister from a previous year's honeymooners, to whom we referred as "the troublesome twosome." Wynona waited for a long time before breaking cover, then brought the herd in on the southeast trail. She had loose stool, which added to the chaotic arrival, as all the youngsters went scrambling to eat the soft feces. Young elephants do this to boost poor digestion. Ahead of the eager dung-eaters, Wynona gave us a crack of her ears as a reminder that she was aware of our presence. She then saw the lions, marched over to their side of the trough, and pushed them back so that her family could drink more peacefully.

This behavior was especially noteworthy, since after a day of basking on their bellies, hind feet straight up in the air, the troublesome twosome had started to work up an appetite, sending everyone at the waterhole into a tizzy. Just prior to Wynona's entry, a small family group had come in with a tiny calf. The young mother was particularly concerned about the lioness sitting in the middle of the clearing, poised to strike. The rest of the group initially showed some concern by fanning out and tucking the little ones in the middle of the group. But when they settled in to drink, some youngsters positioned themselves on the same side of the trough as the lioness and even faced the other direction. This seemed strange, and it was even stranger that the others would not try to scare the lioness away. It wasn't until a very young bull (perhaps a brother) decided to step in and take the lioness to task that the young cow was able to assure her baby's safety.

Wynona, however, paraded around the trough, keeping careful watch over the lions as her herd drank. As was our routine, we collected data on herd spacing and orientation at fifteen-second intervals, as well as data on the vigilant behaviors. After the elephants settled into their drinking session, we ran the acoustic playback experiment for fifteen minutes. In the end, the herd itself didn't respond very dramatically at all, but Wynona was clearly upset.

Later when we played the alarm call back to her herd for a final time that season, I thought for sure she was going to tear into the brush pile where the speaker was hidden and smash it. I almost wish that she had, since it was the end of the season, and would have been justified. After the playback, Wynona put on her chevrons, encouraged the rest of the herd to move away to safety, and marched right over to the speaker to face it. Meanwhile, Luke Skywalker (named for his missing right sword hand, i.e., tusk) happened to be in the vicinity and became agitated, perhaps due to the alarm playback, but more probably because of his excitement at the family group's hurried retreat. He emptied his bowels near the speaker and displayed a huge erection before moving out on the southwest path. Wynona stayed behind, relieving herself right next to Luke's voluminous pile. The releasing of bowels in the elephant world is a measure of excitement, surprise, fear, or agitation, exactly like when a person confronts a potentially lethal encounter.

Wynona's response was completely different from the responses of Slit Ear and Margaret Thatcher. Both took off with their families when they heard the playback, rumbling angrily, posturing as defensively as possible in all directions. It was uncertain why Wynona did not interpret the alarm call as a very dangerous threat but rather as one that she could challenge on her own. For whatever reason, she had a no-nonsense attitude, whether it was that the alarm call was from a herd too distant from her own or that she had gotten used to these annoying noises the year before and simply wanted to put a stop to them.

As our experiments progressed over the month, the patterns of animal movement were more visible due to the late rains. Large numbers of game had only started to move in toward the middle of June, the hoof stock returning in droves since the southern part of the park had now dried up. Large herds of zebra poured in, along with giraffe in such numbers that the perimeter was surrounded with courtship necking and provoked urinations, where the bulls would place their noses up under the tail of a female to stimulate urination, and thus ascertain their hormonal status. In the early morning, the clinking of the eland bull's heels punctuated the predawn hour with the sound of wind chimes. Ostriches danced and hatchlings soon emerged as a result of earlier courtships. Warthog piglets had their debut.

Doves were also returning, much to the thrill of the jackals, which chased them around the perimeter of the pan every day starting at 7:00 in the morning. They were good at catching them, too. I saw a jackal catch a dove, thinking it would suffice for his breakfast, but once he scarfed it down, he was running after another. Even the lioness was feasting on doves and caught one in the equipment bramble. Her brother looked on with disgust as she spit out the feathers.

The lions were not nearly as sexually active this year as they had been the previous season, when we were kept up at night with the noise of loud purring during copulation, then snarled orgasms at their climax. One male in his prime had taken up residence at the waterhole and had to satisfy the seemingly insatiable desires of two lionesses. I had never seen a male so exhausted, as he was forced to perform on demand twenty-four hours a day.

What was remarkable about the sexual behavior was the intensity of the orgasms. As this lion ejaculated, he exhibited an intense growled orgasm, mouth wide open in grimaced seemingly uncontrolled ecstasy, before dismounting and collapsing on the ground, while the female rolled calmly with belly up next to him, the equivalent of having a cigarette, it appeared. When she was satisfied, the next female came purring up, twitching her tail in his pained and utterly exhausted face. He would then force himself up yet again to mount and again attain seeming exaltation. We heard the purring and then the climax almost nonstop throughout the night. The whole camp began to get exhausted for him.

The hyenas had more of a presence this year, and their numbers increased throughout the month. I love the sound of the hyena at night, but there had not been a den near Mushara in previous years. So I was excited to see that in fact there were two dens, one to the east and one to the west. They'd come to drink in pairs, and sometimes in groups of as many as five to eight, all lapping noisily, a frenetic, high-pitched, almost paranoid sound compared to the lazy, confident tongue of the lion.

Occasionally, the dens would clash at the waterhole, resulting in some pretty nasty skirmishes. We had been following what we thought might have been the aftermath of one of these fights through the decline of an unfortunate young male hyena. He appeared to have a compound fracture in his left hind foot. One day, he practically dragged himself into the clearing, then finally and painfully sat down in the pan and lay there for hours. He looked near death. In the afternoon, he got up and slowly hobbled halfway into the clearing but could go no farther. He lay down, and with the exception of a twitch of the tail, he remained immobilized until the middle of the night when a pride of lions came in and surrounded him.

My worst fears for him were realized as the youngsters batted at him with their paws and he howled in terror. Finally, a big black-maned male came over and dispatched with him with one shake of the head, then all interest in him was extinguished. Silence filled the rest of the night, except for the heavy thudding of large cub paws hitting the ground as they took to batting at each other instead.

We went out the next day to check on the hyena and saw the puncture wounds in the head and neck from the final blow, then saw that the mortal wound in his foot was the result of a toe being completely bitten off. Colleen, having been a vet tech in her previous life, inspected the severed toe with interest. She offered me a glove to do the same, but I declined. Still feeling the pain of his final moments of terror, I had no stomach for touching him. I am not sure how I had come to be so sensitive, but perhaps what I feared most was rekindling some of my own pain. Grief has a way of shifting to the forefront with a mind of its own.

In the middle of that night, the rest of the den arrived on the scene. Seven or eight pack members stood around their dead mate as another stood 50 yards away at the waterhole, walking slowly back and forth, giving his slow signature whooping call, only it was so slow and so purposeful—mournful almost—that it was as if he were giving a eulogy. Back and forth he walked and whooped, as the others stood over their den mate. "One man down," he mourned. "One man down." He paced back and forth as if he were questioning the senselessness of death, akin to Marcus Aurelius pondering the honor of Julius Caesar's betrayer, Brutus. And each time he whooped, there was a distant call returned from the competing clan, as if to taunt and remind them that the war was not over. I couldn't help but think that man was not above nature in his own turf battles.

OVER TIME, it became clear that bulls were not particularly perturbed by our alarm playbacks. This made sense, since they are not particularly perturbed by the presence of lions. We had witnessed this many times, most recently with the "gentlemen's club" visit just after the troublesome twosome had killed a young gemsbok that was stuck in the trough. We had thought we had seen the last of the troublesome twosome after the young male had got a serious hiding from two older males a few days before. But they returned, the young male nursing a seeping wound on his right leg and another wound on his back. His sister was suffering from a warthog tusk wound to her left hip. The young male lion was basking on top of the bunker as the first members of the gentlemen's club strolled in, not bothering to chase him away. Obviously, they had forgotten that lions could be a lethal menace from their time in the breeding herd.

With fifty-nine bulls identified the year before, we managed to add twenty new profiles. One of our goals this year had been to focus on their relationships and to collect fecal samples to test hormone levels within the dominance hierarchy we were slowly building. We were interested in cortisol levels as an indication of stress, while testosterone levels would specifically serve to confirm musth. We also planned to look at relatedness within the hierarchy to see if it conferred any status benefit within the group.

By midseason, we had sketched in a veritable soap opera of who beats whom, whom no one dares mess with, and who would suck up in order to stay out of trouble. Recently, bulls we'd named Timothy and Jack Nicholson were often at it, the shoving continuing throughout their entire visit, as each strove to hold his position at the head of the trough, displacing younger bulls in the process. The only break in this pattern occurred when Jeff started a shoving match with Jack, which was a lot more affiliative in nature, with Jeff putting his trunk and head over Jack. This was no doubt highly irritating to Timothy, as it was clear that something was up between Timothy and Jeff that day. Timothy gave Jeff a huge margin when he approached the waterhole, sidling up next to Luke Skywalker, after Luke had displaced Jeff earlier.

Meanwhile, Timothy placed his trunk in Luke's mouth, and they stood, bodies touching as they drank. Luke put his ear over Timothy's face and rubbed back and forth as Timothy leaned in, the closest thing to a manly hug in elephant body language. This must have been a great consolation to Timothy. He had been having trouble lately within the gentlemen's club, often arriving just before or after the departure of the group or even visiting on off days and looking rather beaten up. He had white scar marks all over his face after a midseason drawn-out battle with Prince Charles. Charles chased Timothy around the clearing for several kilometers after Timothy clocked him one at the waterhole, seemingly as a preemptive strike.

In the process of getting to know the bulls, we also started to notice the preadolescents within families, as it was only a matter of time when they would be off on their own. We wanted to understand as much of their relationship history as possible before they ventured off. Once identified, we noted the other young bulls they spent time with while they were still part of the herd, which bulls they frequented when their families met up with others at the waterhole, and whether they reached out to older bulls that were present while their family was drinking.

It's funny to see young bulls jousting within the herd while at the waterhole, whereas the cows are all business: get in, drink, get out. Other researchers have described the cows as being very affectionate, to the extent of appearing emotional with each other, particularly in the case of greeting ceremonies. But at Mushara, for whatever reason, these elaborate ceremonies do not occur, and the females take a no-nonsense approach to their waterhole time.

I finally did get to see one of these greeting ceremonies while I was driving back from a supply run. It actually brought a tear to my eye. I was relieved to see that the elephants were not so stressed that they failed to engage in these special rituals. I was just passing Tsumcor, about a half hour south of Mushara, when I accidentally encountered a cow that I thought was Margaret Thatcher. She came at me, the late afternoon sun glowing on her face towering above me and her ears fully extended. Her mock charge sent me a few meters up the road, but I decided to return to spend more time with her. At Mushara, we were operating in such confined quarters these days that we never got to be close to the families, but here I was, up close and in broad daylight. The cow stood browsing on an acacia branch, her whole body in the road, facing the other way but still watching me. Elephants have an uncanny ability to see with their back to you.

It appeared that my presence was tolerable at this distance, so I turned around and started taking pictures. As I snapped away and tried to look for distinguishing ear features, I noticed that her collar wasn't quite right. Margaret Thatcher has a satellite collar, the satellite portion on the top and a weight with a radio collar on the bottom. This was a very old satellite collar, the weight and radio collar long gone, the belting threadbare, and without its weight, the satellite portion hung below. This was not Margaret Thatcher after all. I noticed a large knob on her trunk. This was Knob Nose, the elephant that friends had tracked almost eight years ago. I had yet to see her in this region, despite this being within her home range. And she had never made her way up north to Mushara during our field season.

Realizing that she was not Margaret Thatcher, I felt safer and remained close as she passed my truck and headed up the road. While she was still standing in front of my truck, another large cow suddenly emerged from the bush on the opposite side of the road and approached her. The two cows stood there, head to head until another came storming out, roaring, ears out, heading straight for Knob Nose. I quickly regretted keeping so close, since this new, large elephant could have backed Knob Nose straight into me. I was trapped. Knowing these elephants were used to vehicles on the main road and that there was rarely an incident with an elephant crushing a tourist vehicle there, I just sat between them, cringing. Then I realized that I was sitting right in the middle of a greeting ceremony.

All those years of living in the Caprivi, where greeting ceremonies occurred every few nights in the dry season on the floodplain in front of our house, I had not been able to get close enough to witness the behavior that coincided with the bizarre roaring, screaming, and trumpeting. Now, here, three cows rumbled, urinated, defecated, and fanned their ears in an excited greeting. Then all three turned to face west in a line, ears out, trunks to mouths, before dropping their trunks flat on the ground in a rigid moment of silence, and finally, they slowly sauntered off together.

Then I realized that the greeter was Donut, another cow that my friends had been tracking. And she had a new baby. I was so close that I had to take pictures through the windshield because I couldn't stick my head out far enough and still be ready to make a quick retreat.

What struck me most about this ceremony was their almost trancelike state. The ritual was not unlike the strange state that flamingos enter when engaging in a group display just prior to the mating season. In these displays, several thousand individuals march in a ludicrous posture, knees chest high, left for several yards, then turning and marching right, all in a line. It is a remarkable ritual.

And yet here was an incredibly intelligent animal engaging in what appeared to be a similar display. This greeting was markedly different from the types of greetings in which the males engage, where the ritual of trunk to mouth is preceded by a sizing-up of the other bull's mood. The bulls don't appear to be in a trance when they engage in these greetings. But what was going on with the cows? Was this ecstatic trance some kind of hormonal trigger? And why urinate and defecate spontaneously as if they were not in control of their bodies? And why the screaming at such close contact? Could it be an innate response, something programmed, perhaps to prime hormones, to facilitate bond reinforcement?

Back at camp, I retreated to my tent early after relaying my exciting tale. Unfortunately, once the adrenaline wore off, I was bedridden again. For the next day's adventures with the gentlemen's club, at least I was able to observe the soap opera from my tent.

By midday, we all had to duck and cover when a large dust devil whirled through camp. The size of these tiny tornadoes could be less than three feet in diameter or more than a hundred feet wide and tall. Leaves, dust, and sand caught up in the mini-twister would speed through camp and leave a layer of sand all over the equipment and notebooks, throughout the kitchen, and in our sleeping bags if our tent wasn't zipped up. Since I was still in mine, I quickly zipped myself in to avoid a dusting.

By late afternoon, I was feeling better and positioned myself in the tower to film Slit Ear and another herd barreling into the clearing together, trumpeting, screaming in their inexplicably frantic approach before settling in at the waterhole. The shot was perfect, the setting sun just in the frame with the herd as it passed, silhouetting them, the whole scene covered in a red backdrop. This was the herd that I needed to study further in order to determine once and for all whether various groups were habituating to my alarm call or whether it was now a herd-specific response. Since Slit Ear arrived with another herd, it wasn't going to be a good comparison, so we spent our efforts looking for distinguishing features with which to identify her and her family better in the future. It was clear that Slit Ear would have a calf by next season, no doubt her first, as she barely looked old enough to mate.

In the meantime, a musth bull came in and began to create chaos among the herds. We hadn't identified this bull yet, so he got the temporary title "Musth Bull." While we waited for the herds to leave so we could collect fecal samples, we watched them dust themselves in the blood-red backdrop of the setting sun. As they dusted, the little ones dropped to the ground and rolled, feet in the air, while the mothers kept a close watch and dusted themselves. The dust created a path straight into the setting sun. It was as if the great western elephant highway were lit up to guide the way.

The herds frolicked and dusted for some time in small family groups until all forty-five decided to head out and a new herd approached from the northwest. The musth bull returned to inspect the new females, placing his trunk in intimate places. Luke and another bull we called Billy the Kid stood clear of the musth bull but also inspected the females. There was an occasional wail and trumpet from a young male, no doubt as he was put in his place by one of the adult cows.

AT SOME POINT in the middle of that night, I awoke to the sound of water trickling from elephant trunks. It was Wynona and her herd; apparently, they hadn't had their fill after the late afternoon playback disturbance. They were standing in a line, drinking quietly, only the noise of the water marking their presence. I watched and counted with the light of a half moon overhead. The shadow of Wynona's W cutout on her left ear was plainly visible as she turned to face sideways during her vigilant scans of the horizon. One of these was timed with the arrival of Gakulu, a bull who seemed to like to visit the girls in the night. Some of the younger cows greeted him with a trunk to his mouth.

It was clear that Gakulu had a special rapport with the females. If any of these cows were in estrus, they'd surely present to him by backing up and allowing him to inspect their vulva. None of them went this far, but he did do a spot check, just to be sure. But not Wynona. She was busy patrolling, walking up and down the ranks of the family, making sure all was in order. If I didn't know better, I might have mistaken her for Margaret Thatcher because of the way she was behaving.

Later I learned from the hormone results of her fecal samples that she had very high testosterone levels for a cow. Another researcher had suggested that an increase in testosterone levels might explain her aggressive behavior, but this was the first time we had considered the concept. Not knowing where it would lead, we decided that the following year we would collect hormone samples from different members of several family groups to see what else we could discover about this phenomenon.

What I really wondered was what defensive maneuvers Wynona would exhibit in the next season. Would she finally make her last stand and destroy our playback system once and for all? I savored the thought.

16

THE GENTLEMEN'S CLUB

The elephant's a gentleman.

—RUDYARD KIPLING, *Oonts*

"INCOMING FROM THE SOUTHEAST!" I always felt like Captain Ahab when I said that, hanging from the ladder of the tower with one hand and looking through my binoculars, as if hanging from the crow's nest of a great ship as giant white elephants approached. The crew grabbed for extra layers of clothing and red headlamps and scrambled up to the lower platform, settling into position to video the herd and prepare for an experiment.

The sun had already set, and the first sliver of the waxing moon gleamed as a large family group came spilling into the afterglow, attended by Gray, a musth bull. I counted individuals as they passed the tower, looking for identifiable features to define the family. A young spunky bull gave us a head shake as he passed, his extended ears revealing a deep V-shaped cut on the middle of his left ear. The presence of this character gave me a hint that it was Slit Ear's herd, which should have been obvious, as Gray had been following them for the past week. Musth bulls are thought to spend much of their time following family groups in pursuit of females in estrus. There must have been a female coming into estrus in Slit Ear's herd, as more and more bulls displayed interest in them, particularly Gray and some of the others in musth.

I moved up to the top of the tower, started the tape recorder, and did some video closeups of the herd while there was still some light to capture more ear details from the older cows. We were planning to conduct a control experiment to document normal behaviors exhibited when no alarm call was played.

But just as Slit Ear's herd settled in to drink, all hell broke loose in the clearing. Herds were arriving from every direction—from the southeast, southwest, and northwest paths. One of the herds from the southwest burst into the clearing trumpeting and roaring, then stopped short of the waterhole as if the elephants had suddenly become aware that the waterhole was already occupied. They stood in a massive clump of thirty or so, an extended family group, sniffing with trunks raised toward the waterhole. One of the cows shook her head, frustrated that they would not be given right of passage. They stood and waited as Slit Ear's herd drank, Gray patrolling up and down, inspecting females and displacing the young bulls with a head toss, scattering them to a safer distance. Some of these same young bulls were also given a sharp jab or two by the adult cows, and in one case, the threat of an approaching dominant female sent a young bull bellowing off.

It was only a matter of time before that young bull would be tossed out on his ear. We were starting to recognize a few of these fellows and hoped to be able to watch their transition into bull society when that time came. Because elephants are very much creatures of habit and social bonding, we assumed that groups of bulls most likely cohered based on relationships that began at a young age within their families, as they grew up sparring with brothers and cousins. If that were the case, then entering bull society would be a lot less stressful than having to start from scratch. We were hoping that the results of our genetic analysis would shed light on this issue.

Earlier that day, we kept ourselves entertained when Greg (the don) returned on his regular every other day schedule, entering the arena with his merry band of cronies, head held high, ears out, and strutting his stuff. The don only exposed his identity somewhere within the last 50-yard distance to the water. After that, everything had already been previously negotiated. There were only a few altercations involving Greg at the waterhole, and so far, these had only been challenges from musth bulls, either from residents or passersby, but they always yielded to the don in the end.

At 2:00 P.M., thirteen bulls appeared on the horizon in one long stream, an amazing sight, their large heads swinging back and forth in procession, the younger ones interspersed between them, spacing themselves a body length or two apart. A herd of bulls has a different dynamic than the entry of a family group, where heights graduate more subtly into different age classes. Families constantly clump and file out when comfortable, then again clump and then go single file, depending on how confident they feel about their safety at any moment in the waterhole arena. A bull entry is much more ordered—that is, until the midranking bulls hit the waterhole. Then there is some jockeying.

The parade had begun with Jeff earlier that day. He had arrived early enough to drink in peace before Luke strolled in and displaced him at the head of the trough by a wide margin. Jeff retreated and sulked at the far end of the pan until reinforcements arrived. Then in strode Torn Trunk and David the Lion Slayer, followed by Vincent Van Gogh, a new half-sized bull, named for the two V-shaped slits out of his right ear; Timothy brought up the rear. Finally, Greg arrived, head held up, ears out, mouth wide open, and marching straight for the head of the trough. All gave way without hesitation, except for Torn Trunk, who was starting to show signs of going into musth, as his penis sheath had started to get crusty. It was as if he couldn't help but face Greg head on as he approached. This riled Greg even more, who came straight for him, until at the last second, Torn Trunk backed down, turning his head to the side in submission. The don would not be challenged that day.

Meanwhile, most of the bulls were on their best behavior, exhibiting gentle gestures of affection toward one another by body rubs, trunks to mouths, ears over faces, and trunks on backs. However, a few ongoing spats did not seem like they were going to be resolved that day. And after a full five-hour visit, the bulls finally pushed off, all the subordinates following Greg out in a long line.

Having noted who defecated, when and where, we headed out after the bulls departed, dung map in hand, on our late afternoon dung safari. We collected fecal samples, measured hind footprints, and took photos of both hind feet, the larger of which measured around 23 inches, and the shorter, stockier bulls had foot measurements of around 22 inches. Curiously, these measurements did not seem to match the idea that bulls grow throughout their lives, their hind feet correlating with a continuously increasing shoulder height.

When conferring with other elephant colleagues, it became evident that these desert-dwelling bulls had much larger feet than those measured elsewhere, but to us, the measurements did not seem to scale with age. We planned to solve this problem in the following season by using an inclinometer, a tool that looks like a protractor that can measure the angle of an object to determine its height. Comparing shoulder height to hind foot measurements, we could have a better idea of what a good measure of age would be for this population. If a photograph was taken at the head of the trough (a known distance) using a fixed focal length (a 400-millimeter camera lens), pixel density could be counted and compared between photographs to get an accurate shoulder height. Ideally, we would have data from teeth from mortalities data or an immobilization event to compare to the shoulder, but since we didn't have that luxury, hind foot length and shoulder heights would have to do.

We were most excited to find that the patterning of cracks and wearing in the footpad was different for each elephant, so we started a catalog of Mushara bulls based on footprints to supplement obvious physical features. These data supported the most important data set, the behaviors, giving us information about relationships, whether friendly, such as trunk-to-mouth greetings, or aggressive, such as shoving, or a tusk to the buttock to displace a lower-ranking bull from the head of the trough. Based on these events, we were able to compile a hierarchy of dominance, with Greg, the don, sitting on top. He appeared to have the most trouble with Kevin, the third-ranking bull, and Mike, second in command, didn't have an aggressive bone in his body.

It just so happened that, along with Kevin's bald tail, he also had bald feet, as if he really had been up to some serious back-alley scuffles. Perhaps the worn footpad not only gave away the maker, but was an insight into his character as well. Greg had a worn pattern on the right side of the right heel, but nothing more than that. As would be fitting of a don, he did not have to put much physical effort into his threats, and so didn't show much wear and tear on his feet. Of course there would be other, natural reasons for wear and tear on the feet, not unlike the wearing away of the sole of a shoe.

As we were out collecting data, Willie Nelson sauntered in, no doubt avoiding the gentlemen's club. Willie was named for his character, despite his short, stocky stature. He appeared to be an older bull due to the size of his head, the thickness of his tusks, and the quality of his wrinkled skin, looking delicate and loose, like an old man. He had a cool, calm demeanor and a very long, thick tail like his namesake, and a worn, distinctive left ear with a big cutout on top, followed by three dangling fingers of tissue that were quite dramatic when he held out his ears. He was well respected within the community but chose not to interact with the don. I couldn't figure out why Willie was such a loner that year. There was still no sign of his best buddies, Orville and Abe, after three weeks of observations. They had been almost inseparable the previous year.

We watched on the perimeter of the clearing as Willie approached the water. We had gone out to collect dung from five members of the gentlemen's club, and one and possibly a second musth bull, including Torn Trunk. It was very interesting to see the progression of how Greg dealt with bulls going into musth, in this case, taking Kevin to task. Kevin, by himself, was a formidable presence and the third-ranking bull within the club, but he buckled under the threat of Greg, who circled clear around the pan to challenge Kevin upon Kevin's late arrival, resulting in an all-out clash of the titans. Greg's ears folded in extreme aggression as he knocked Kevin back with a tusk-on-tusk collision. In the heat of the combat, Kevin had managed to put the bunker between himself and Greg, so they kept circling the bunker, squarely facing each other, Greg swinging his foot at him in frustration until finally the standoff was broken and Kevin retreated to the perimeter.

Although mostly subtle, Greg must have been kicking a lot of butt behind the scenes to get some of the reactions that we had seen at the waterhole, where he calmly commanded his reign over the trough. Normally extremely cordial, it was only the rare clash that let us know of Greg's true potential.

Dominant bulls are thought to go into musth only during the rainy season, when they would encounter the largest number of females in estrus, but we had seen Greg in musth the previous July, so perhaps he was just starting to cycle into this aggressive hormonal state now as we reached the last third of June. But he showed no physical signs of musth. It was clear, however, that he was intolerant of other bulls in musth and was sometimes violent with them.

Willie stood and drank, watching us for some time as we collected fecal samples from Vincent Van Gogh, and from Kevin,

Torn Trunk, and David the Lion Slayer, who was half to barely three-quarters in size and for whatever reason had clearly gained favor with Greg and sucked up whenever possible after a potential threat by Jeff or another mid-ranking bull such as Jack Nicholson. He was like the little dweeb on the playground in second grade who would cause trouble and then run for cover behind an older cousin and renowned bruiser. He'd give Jeff or Jack a poke with his tusk, or see that a blow was coming his way, and he'd run to the head of the trough, slide in under Greg's ear—Greg seemingly unperturbed by David's ventures—and drink from the head of the trough, chest held out under the shirttails of the don, beaming with inexorable pride. He knew he had it made, just as long as the don was within reach.

By the time we finished collecting dung and measuring footprints, Kevin had strolled back in, conveniently providing us with a late-stage musth sample. He and Willie engaged in a delicate tap dance around the prime position at the water trough, Willie reluctantly giving way to this musthy menace. After a long drink, they gently sparred as they left the waterhole together along the southwest elephant highway. It was as if Kevin returned to spend time with Willie, who appeared to be trying to keep a low profile and was reluctant to partake in Kevin's persistent sparring. This wasn't the first time that Kevin seemed to seek out Willie for a good solo spar.

Kevin was just coming out of musth, still temporal staining but no urine dribbling. He still exhibited musth behaviors, with classic, exaggerated trunk curls over the head and then his signature dragging about two feet of his trunk on the ground. He would also swing it in Willie's direction, who in turn contracted his trunk so that it fell just above his knees, pushing back, as if hoping Kevin would stay put and stop approaching him. The two-step went back and forth, a trunk push and two steps back, then another trunk shove with one trunk placed over the other's head and back two steps in the other direction. And so it continued for some time before Kevin decided to give up and head out into the setting sun. Willie hung back at the perimeter of the clearing. I had a feeling I'd be seeing him again later that night, as he often provided me with late-night entertainment.

When we got back to camp, it was just lighter than twilight as I scaled the ladder to see if any family groups might have approached within sight while we were on our dung safari. That's when I caught sight of Slit Ear's herd and we set our tower operation in motion.

At this point, a third herd had arrived, scattering both the herd in waiting and Slit Ear's group, and causing a scuffle among the bulls and anyone caught in the middle. Sure enough, it was Margaret Thatcher parading in, head held high, tusks like scimitars. We took note of the young bulls in her family that looked about the right age for their eventual rite of passage. I wondered what these bulls might be like as adults. Would they have learned a thing or two from Margaret about how to stay on top? Would they become the "bullies" of the bull world? This got me thinking about the young bulls in the other families we were following. If mothers' characters had an influence at all on their future status, then perhaps Wynona's sons would be benevolent dictators. I couldn't help but wonder what Willie Nelson's mother had been like. Would she have been upset that he turned out to be a cowboy?

17

THE SAILS OF MUSHARA

Do what you can, with what you have, where you are.

—THEODORE ROOSEVELT

EARLY ONE MORNING, IN 2005, I woke to a strange thudding sound approaching the camp from the south. It was as if Frankenstein's monster were approaching with leaden feet. There was a thud, then silence while a foot swung forward, then another thud. It was almost two in the morning and pitch black out. I had never heard a sound like this before. I reached for the night vision, turned it on, and peeked out the side window of my tent to take a look.

Sometimes rhinos sounded rather clumsy, but their footfalls didn't have such resonance. I thought maybe it was Kai, the white rhino, reintroduced to the wild by a German zoo. He often strolled in late at night, trailing close behind a female black rhino and her calf, or behind a frustrated bull not wanting Kai to get so close. White Rhinos are gregarious; blacks are solitary. Kai was having a hard time fitting in. He chose to spend his time here rather than in the south of the park with the other white rhinos, but it was hard not to feel sorry for him in his confusion.

I saw the offending beast as he emerged from the bush, heading straight for camp. It wasn't Kai. In the grainy green viewfinder, a large musth bull came into view, swinging his trunk out in front of him like a fire hose, then dragging it on the ground between his legs, all the while clump, clump, clumping toward us as if each foot weighed a thousand pounds. Elephants were usually light on their feet, so this behavior was unusual. Equally odd, he brought his hind legs forward, his hind feet falling well in front of his front footprints, where normally each would fall just within the last imprint of the front.

I tracked him in my scope as he thudded his way toward me, but I had to keep taking my eye away from the viewer to see how close he was really getting. As he approached, it was harder to keep my perspective. The bright illumination of the night vision meant that what little natural ability I had to see in the dark was ruined. I had no choice but to watch the flickering, fuzzy version of reality.

I was glad that the others were asleep. This was the closest an elephant had ever dared to get to our camp, and at the rate he was dribbling urine, his testosterone levels were likely at their peak. Some of my assistants were comforted by the imaginary perimeter that the elephants defined for our camp, usually at about 100 feet or so from the boma cloth, but this guy thought nothing of skimming the wire stays of Camp Mushara.

He had no interest in us, however, and went straight to the water. Bulls in musth appear to have to drink more often because of the amount of water lost from their urine dribbling. Musth was becoming a more and more curious phenomenon to me, particularly after recent interactions I had witnessed. A week prior, Greg (the don) had finally confronted Torn Trunk, as Torn Trunk was then fully in musth and dribbling urine at an alarming rate. Greg cornered him for over an hour, penis fully extended, ears out and threatening. It was as if Greg were saying in his Marlon Brando voice, "This musth, it's an ugly word."

Torn Trunk, by no means an insignificant bull in age and size, stood almost apologetically next to Greg, dribbling madly from his temporal glands and erect penis, as if trying to shut himself down, to pretend his musth signaling wasn't really occurring. Was Greg trying to suppress Torn Trunk from going into musth? Cases of suppression had been documented in which the presence of older bulls influenced the age at which younger bulls would first go into musth—but not like this, between older bulls of equivalent size.

I discovered from our hormone data that in fact Torn Trunk's testosterone levels were much lower than they should have been, given his outward signs of musth. This made me wonder whether the older bulls were able to manage this hormonal state to the point of dissociating cause from effect, testosterone from musth. Or was Torn Trunk in the process of shutting down musth so that he wouldn't be tossed out of the group? Either way, it was an extraordinary finding.

Greg's testosterone levels shot up within this three-day period, perhaps needing that extra surge to suppress Torn Trunk, who had formed a splinter group and kept his distance from the gentlemen's club for the remainder of the week, presumably to avoid any further trouble with Greg. What energy Torn Trunk did not expend in testosterone at that time was made up in cortisol, a peak in this hormone indicating high stress levels, most likely from his encounter with Greg.

I could see a whole new study unraveling before me. After watching this musth bull drink a while longer, absorbing the environs, I looked across at the bunker just beyond where the bull was standing. It was hard to imagine that fourteen years had gone by since I first came to this site, studying elephants from that dank box. To be honest, I still had a certain nostalgia for the bunker. Living right in the midst of all the action had been a fantastic experience, but I had to admit that the tower, the camp, and the roof tent setup that had evolved over time was much more practical. For the discerning bush camper, you might even say it was luxurious.

I barely had to lift my head off the pillow in order to see the waterhole and its late-night visitors. Not only could I appreciate the night sounds of Mushara, but with our sleeping arrangements, I was easily able to maintain an operational position throughout the night, lying half awake, finger poised on the record button.

When it had just been me in the hide, I was limited in the types of questions I could ask; what I could accomplish had to be tailored accordingly. Still, the bunker days of just me, my gas cooker for tea and pea soup, and the flatulent bulls for company occupy a fond place in my memory. An enthusiasm as raw as the surroundings sustained me in those days. I never could have made the recordings that I needed to prove my hypothesis had I not spent considerable time in that cement box.

Over time, however, it became clear that I needed more eyes and more minds to accomplish my goal. After three seasons with a team, I finally became used to sharing the tranquility of this place, the close quarters, the idiosyncrasies, the little noises that people need to make. But the compromises were amply compensated.

The idea that elephants might use the ground as a sounding board had initially baffled geophysicists because most of them had taken pains to eliminate the ground surface waves that interested me. Because these waves were considered noise, they had been filtered out of data sets, and there wasn't that much information about how far they traveled.

Little did I know that this idea that began as a niggling curiosity in the field in Namibia would then take me to the bayous of Texas, the Nevada desert, southern India, northern Zimbabwe, the Oakland Zoo, and then back to the scrub desert of Etosha pan to find an answer. I now know that those tiptoeing stethoscopes were being put to good use.

The waning moon appeared on the horizon. I stayed up way too late again, savoring the last few nights in the field. It was starting to get colder as July set in, and I was grateful that we wouldn't be experiencing the full force of the frigid July nights. I lay there thinking how much this new camp had become a part of me. My challenge as a scientist wasn't that there weren't enough interesting questions to ask but that I would have to remain focused on the first question and find the answer while being open to the next question presenting itself. I never expected to learn enough geophysics to show that elephants could detect their own vocalizations through the ground. And now, having solved that mystery, a new mystery arose, requiring another toolbox and a different set of eyes.

I stared at the Southern Cross in its final stages of migration across the sky, looking every bit like a kite with its string severed, sinking sideways into the horizon. I had named a bull in honor of this constellation, "Daniel the Star Gazer." He appeared on the eastern horizon that day, my loyal friend who had spent the most time keeping me company through my long bunker days. Why hadn't I looked for him, expected him to show up? I had completely put him out of my mind, somehow thinking that those bulls, just as those days, were long gone.

I was suddenly and humbly reminded that as capricious as nature could be, it was also inexorable. The bulls were a permanent fixture of the landscape, just like the Southern Cross that adorned the tower for much of the night. I was the only presence that came and went here, just a speck under a constant sky.

As I lay in my tent, I felt the rotation of the earth, the changing of the seasons, and the certainty that when the time was right, the elephants would disperse once again. But they would always return. I hoped that I, too, could be certain in my constancy with this place, but I knew down inside that I might not have that luxury forever. Every year that I returned to Namibia, as excited as I was to be with the elephants again, I worried about their future. How much longer would I be able to study elephants in environments where their societies are still intact? Would the war between farmers and elephants escalate to the point of irreparable damage to one or both societies? Even though things looked favorable for elephants in Namibia because of the government's progressive conservancy initiatives, what would be their fate elsewhere? Reports of killings by both elephants and farmers were flooding in from all over Asia and some parts of Africa on a daily basis. And recently, it has been suggested that these murderous battles may be caused or aggravated by a form of post-traumatic stress disorder (PTSD): young elephants may have witnessed the death of a family member, and either through loss of a mentor or due to trauma, or both, these elephants may be engaging in unusually violent acts as a way of seeking revenge.

This thought haunted me as I remembered the last time I had seen Daniel. It was New Year's Eve 1997. The clouds reeled as lightning whipped the earth, when suddenly he appeared to drop out of an infinite purple sky, as if born from an electric tongue. His form broke from the horizon, and after some time, loomed large in the foreground. Having had a good long drink, he slowly took a mud bath, then came over to pay his respects. It was unusual for me to be there in that season, and I wanted to think that he seemed pleasantly surprised to see me. I was certainly thrilled to see him, the only elephant I had seen during that three-day sweltering stay to measure thunder in the ground.

He came over to me, those wide-splayed mammoth-like tusks all caked in mud. He had a gentle yet robust demeanor, as if he had been around the block so many times that he didn't need to posture. He stood for a few minutes to take me in, playing with the bottom of his trunk, holding it just off the ground, twirling the tip, sniffing in my direction. He looked away as if to pretend that he wasn't really trying to take me in, but he was. He held his head up toward me for one last look and then wandered off, swallowed up by a giant, wavering horizon.

I AWOKE THE FOLLOWING MORNING to a pawing sound at the back of the kitchen area. I had heard lions around dawn, so I asked Colleen to be sure to climb the ladder to have a look out back before going out of camp in the early morning. I hadn't heard from her since, then dozed off again. In a moment of sick fantasy, I worried that this pawing sound was one of the troublesome twos batting at our wall and that the other had somehow quietly made off with Colleen. I quickly jumped down from my tent and climbed a few steps up the tower ladder to get a better view.

Twang. Twing. Realizing that the noise I was hearing was the stabilizing wires being cut from the boma cloth, I ran outside. There was Colleen, running down her checklist, one wire at a time. I knew that she was just going through her own process of leaving, gathering up both mentally and physically, but the problem here was that she was taking down the fence two days early. I suggested that there were lots of other things that could be done and that the perimeter would be the last thing we took down. She smiled bashfully and admitted her real motives. She was worried that there wouldn't be enough time to take data when the gentlemen's club arrived. They were expected to return on the third, the day we were supposed to pack up the camp, and she was only trying to find more time to spend with the bulls.

Since all its wire stays along the western wall were cut, whenever the wind came up, the camp appeared to have a spinnaker billowing in the direction of the western elephant highway. During our last dung collection safari, we laughed at how, from a distance and with each gust of wind, the camp looked like a clumsy ship about to set sail. It was as if the whole operation would be ready to launch after the gentlemen's club visit, following them on their way to some unknown, distant place.

On the last night, when the rest of the stays around the whole camp had been cut, I lay awake listening to the gently luffing sails of the SS *Mushara*. After the silence of the early morning hours, the flapping began again, the gusts causing the ash in the fire to twirl. The smell of the ash in the open frigid air brought on a moment of nostalgia for the Caprivi in the dry season, as crisp early mornings were often covered with a smoky blanket of dense air. I reached out of my bedroll, placed the kettle onto the remaining coals, and stared at the spout until I saw the first few wisps of steam emerge.

We all then got up in the predawn light to finish breaking down and packing up. While a herd of eland drank at first light, Johannes hauled down the sails. Then we rolled them and Johannes stuffed them away as the sated eland headed off toward a rising red sun.

Eventually, we squeezed the last remaining equipment into the trucks and I took one final look around. The place looked so deserted without our boma cloth, the tower so stark and naked without its second platform and ladders, and the bunker as bleak as ever. I kicked my foot around in the sand just to make sure we didn't leave any loose wires buried. I was stalling. Seeing the gas cooker packed just inside the tailgate, I couldn't resist thinking that all I'd need was the stove, a water jug, some tea bags, dried soup, and my sleeping bag . . . maybe Johannes could come back for me in a week.

ACKNOWLEDGMENTS

ONLY A LIFETIME OF INFLUENCES make a book such as this possible. From my first educational influences, to the teams of collaborators from across the globe that helped me bring this work to life, there are so many to whom I am indebted.

In particular, I would like to give thanks to my parents for imbuing me with the confidence to believe that I could be and achieve anything that I set my mind to. And to my husband, Tim Rodwell, for his willingness to join in my many projects, and for making them infinitely better, as well as his patience and proficiency as chief technology officer in the field.

I am grateful to my grammar-school science teacher, Mrs. Mcgurty, high school science and math teachers, Mary Jane Rothelin, Sister Lawrence, and Miss DiGiacomo for instilling the desire and confidence in me to go further in science. Also to my ecology professor at Fairfield University for assigning Aldo Leopold's *Sand County Almanac,* and to my organic chemistry professor, Dr. Polito, for inspiring me to get back on the mountain after having lost my footing. To art professors, Philip Eliasoph, Peter Gish, and Rick Mills for helping me to recognize the beauty in science.

I am indebted also to a childhood friend, Joe Walsh, who first invited me to be a field assistant on Guana Island, an experience that changed my life and directed my future on a path that I had only dreamed was possible. To the proprietors of Guana Island, Henry and Gloria Jarecki, for generously providing young scientists the opportunity to get their feet wet in a unique and spectacular setting. And to Skip Lazell for his dedication to educating young scientists. If it weren't for my introduction to the glue shooting *Nasutitermes* termites and the great schools of anchovies attended by tarpon, I may never have taken pen to paper to describe the behavior of animals.

It is not easy for a young scientist to get started, and it takes a special kind of mentor to take on unproven scholars and give them their first chance. I consider myself very fortunate to have had a life rich in such mentors. I thank Scott Miller, then Chair of Entomology at the Bishop Museum, for introducing me to the world of entomology, and the whole department for taking me under their wing; Ken Kaneshiro, my thesis adviser at the University of Hawaii, for bringing that world to life; as well as Hannelore Hoch and Manfred Ashe, whose enthusiasm for planthoppers was contagious. I am grateful for the support of Sheila Conant, Jim Archie, and Chris Simon, and also M.S. committee member, Rebecca Cann, for her influence on my scientific thinking.

My husband and I arrived in Africa with a passion for field work but little practical experience in large mammal conservation. It took the trust of a few key people to get us going. Our first willing mentor and guide was Leo Braack, former parasitologist at Kruger National Park. The opportunity to work with wild elephants on the Caprivi Elephant Project was a great honor bestowed on us by the government of Namibia through the trust of Malan Lindeque, former head of research at Etosha Ecological Institute, now permanent secretary of the Ministry of Environment and Tourism. Over the course of fifteen years of work in Africa, many people proved vital to our success. I am deeply indebted to those who have continued to support us to this day: Chris Brown, Pauline Lindeque, Wilfred Versveld, Johannes Kapner, and most of all Jo Tagg, whose dedication to conservation and powerful character kept me sane and served as the original inspiration for this book. To Grant Burton and

Marie Holstenson, former managers of Lianshulu Lodge in Mudumu National Park for their friendship and generosity; and in Etosha, Ginger Mauney, Nad Brain, Werner Kilian, and Wynand du Plessis. None of the pivotal seismic experiments would have been possible without the constant support of the MET management staff at Namutoni Ranger Station, in particular Immanuel Kapofi who has always "been there for us." I would also like to thank IRDNC Directors, Garth Owen-Smith and Margaret Jacobson for providing a beacon of light in a sometimes dark and lonely Caprivi. I have especially fond memories of my colleagues and friends from the tumultuous Caprivi days. Matthew Rice, Simon Mayes, Beaven Munali, and Janet Matota, the late Loveness Shiita as well as Barbara Wyckoff-Baird, all of whom made community work in the Caprivi not only possible but inspiring. And to Namibia Nature Foundation founder Douglas Reisner within the Olthaver and List Group, and especially Annatjie du Preez, who took me in as a sister whenever I was in Windhoek.

I have many to thank for guidance during my dissertation research, especially my Ph.D. advisers from UC Davis: Lynette Hart, Bill Hamilton, and collaborating geophysicist, Byron Arnason. Lynette and Byron's enthusiasm and dedication to my research contributed to the initial cracking of the code. I'd also like to thank my third dissertation committee member, Peter Marler, for fruitful discussions and suggestions, as well as Whitlow Au, Peter Narins, Darlene Ketten, and Thomas Hildebrandt.

I am indebted as well to my postdoctoral mentors at Stanford: Paul Ehrlich, Chris Contag, Alan Schwettman, Simon Klemperer, Robert Sapolsky, and Robert Jackler, and especially to Sunil Puria for always coming up with with a solution to a difficult problem. I am grateful for the intellectual and field support provided by my current postdoc collaborator, Jason Wood. Also, thanks to the creative thinking of Don Greenwood and USGS geophysicist, John Evans. A special thanks to the Contag lab members at Stanford Medical School who took me in like a family member and became my molecular mentors with a generosity of spirit from which I have greatly benefited.

I thank also my Rotary sponsors, Pat Derby and Ed Stewart, and the staff at the Oakland Zoo elephant barn, in particular Colleen Kinzley, whose dedication to elephants has added so much to my understanding of this amazing animal. I am grateful to recent collaborators for their support, advice, and enthusiasm: Joyce Poole, Sam Wasser and his lab, Donna Bouley, and again to Robert Sapolsky, the depth and breadth of whose thinking continues to serve as inspiration.

With every large research project comes a great deal of labor and creative input from many people. I thank the research assistants and volunteers who worked on the elephant field research over the years, in particular: David Shriver, Nora Rojek, Lianna Jarecki, Katie Ekhart, Ben Hart, Megan Wyman, Seth Haines, Roland Gunther, Renee Hoyos, Sarah Partan, Christina Alarcon, Mike Carpenter, Gina Gambertoglio, Quiana Knight, Willie Phan, Roxana Ramos, Jennifer Sands, Cheryl Matthews, Donna Why, Yoshi Hirano, Judy McVeigh, Max Salomon, Dan O'Connell Sr. and Jr., and also the Wilderness Travel volunteers in India.

I thank my most recent writing group for their encouragement and guidance with earlier drafts of this book: Martha Hoopes, Anne Davidson, Rob Sparr, and Brian McCauley. I am also grateful for comments by Christy Brigham, Michael Marchetti, Rebecca Whitney, Lisa Bohannan, Connie Rylance, Joy McCauley, Colleen Kinzley, Lynette Hart, Margaret Jacobsohn, Pauline Lindeque, Jo Tagg, and Tim Rodwell, as well as my dad, Dan O'Connell, who helped me shape the book proposal with his sharp eye and steely pen. I also thank professors Jacob Molyneux, and Marvin Diogenes from the Creative Writing program at Stanford, and Mark Schwartz at the Stanford News Service, as well as authors Al Young and Ian Douglas-Hamilton for their encouragement.

As in science, it is difficult to break into writing. It takes an agent willing to trust and nurture a new writer. And for that I am grateful to my agents, Karen Nazor and John Michel. I have learned a great deal from John and am forever indebted to him for his patience and guidance throughout the process of writing this book. I am also eternally thankful to my editor, Leslie Meredith, and her assistant, Andrew Paulson, at The Free Press, for enabling me to share this work with the world, and for the huge amount of work they have put in to get this book into print.

Technical support was provided by U.S. Geological Survey, PASSCAL, RefTek, Geometrics, Rainbow Electronics, Cetacean Research, SpectraPlus, Noldus and E·A·R in the U.S. and Thompson Radio and SolarTec in Windhoek, Namibia. I thank the elephant facilities where some of the early seismic work was conducted, including: PAWS, Marine World Africa USA, Bucky Steele's Elephant Facility, Wild Things, and the Caldwell Zoo. Also The Elephant Camp, Victoria Falls, Zimbabwe, the Kabini River Lodge, and the former logging camps surrounding Nagarahole Park in Karnataka, India.

I am grateful for the financial support of many government and private research organizations including the Namibian Ministry of Wildlife and Tourism, National Geographic Society, U.S. Fish and Wildlife Service, The European Union, United States Agency for International Development, Namibia Nature Foundation, World Wildlife Fund U.S., Integrated Rural Development and Nature Conservation, Rotary International, Wilderness Travel, National Science Foundation, University of California, Davis, Art of the Wild, Stanford University Bio-X Interdisciplinary Research Award, Morrison Institute of Population and Resource Studies, The Oakland Zoo and paying volunteers through Utopia Scientific, and lastly, the Seaver Institute.

Finally, I thank my brother Dan, a constant source of creative inspiration with whom I never got out of the sand box.

INDEX

Abe (bull elephant), 214
acacia trees, 21, 84, 86–87, 90
acoustic communication, 56
acoustic fat, 48–49, 184
Africa:
  dual aspects of, 23
  history of elephants in, 82
  mirrors and, 142
African elephants, 82
African sleeping sickness, 114
Afrikaans language, 24, 27, 31,
    135–36, 146, 147
Agency for International Development,
    U.S. (USAID), 24
AIDS, 33–34, 99, 129, 172, 186
allomothering, 63
Amarula, 190
Amboseli National Park, 68
American bison, 6
Andara (Catholic mission), 138
Angola, 21, 115, 117, 120, 130, 141,
    172
anthrax, 92–93, 94, 95
anti-aliasing problems, 181–82
anti-Americanism, 24, 27, 28
anti-malarial drugs, 43, 45, 75
Apaches, 7
arthropods, 6
Asia, 222
Asian elephants, 82, 124, 185
auditory cortex, 7

baboons, 80, 88
baby elephants, 40, 196
  learning control of trunks by,
    14–15, 64, 94
  protection of mothers during deliv-
    ery of, 17
  running ability of, 17
  sick and orphaned, 93–95

suckling by, 15
vulnerability to lions of, 15, 16–17
bacteria, 92
Baikea (Rhodesian teak) woodland, 72,
    85–86
beetle larvae, 87
Benguela Stream, 147
Bent Ear (retired matriarch elephant),
    62, 63
Berlin Zoo, 185
Billy the Kid (bull elephant), 207
biltong (dried meat), 115
birds, 185
birth control, 83–84
Bishop Museum, 122
bison, 6
black rhinos, 217
blind mole rats, 6, 78
blue whales, 57
body parts, selling of, 130–32
boma cloth (game capture netting),
    173, 218
bond units, 14, 62, 68, 90, 91
bone-conducted "hearing," 184
border crops, 51
Botswana, 21, 24, 28, 127–28, 141
Botswana Defense Force (BDF), 115,
    130
brain, 7
British Virgin Islands, 121
Broken Ear (matriarch elephant), 1–2,
    4, 8, 11, 60
buffalo, 23, 72, 85, 86, 87, 128, 154
Bukalo chief, 148, 154
bull elephants, 3, 4, 12–14, 71–72, 78,
    83, 171–72, 175
  alarm call responses of, 60–61, 202
  bachelor groups of, 16, 193–94,
    209–16
  cows in estrus protected by, 113

231

bull elephants, (cont'd)
    dominance among, 16, 194, 211–16
    dominance in mating by, 84, 191
    estrus call responses of, 188–89
    hormone levels and, 203
    internal testes of, 84
    Kinzley's identification book for,
        189, 194, 196, 203
    as most powerful bush animal, 13
    quantities of water drunk by, 78
    searching for cows in estrus by, 8,
        200
    size of hind feet of, 212
    tree-shaking by, 74
    trunk-in-mouth greetings between,
        16, 63, 194, 203, 206, 208
    vasectomies performed on, 84
    vocalizations made at waterholes
        by, 13
    see also musth; young bull elephants
bunching, 68
bush babies (night apes), 131
Bushmen, 128, 140, 141, 187
"buttkickers," 188
Byron (geophysicist), 184

calcrete, 12
Capetown, 147–48
Caprivi:
    author's excursions to villages of,
        40–52
    dung and vegetation transects con-
        ducted in, 80–90
    East, 31
    government postings to, 32–33
    war in, 172
    see also West Caprivi Game Park
Caprivi Liberation Army, 128–29
capsicum (red chilies), 51
catfish, 166–67
cats, 185
cattle, 127, 155
census studies, 24–25, 85, 114–17
Center for Earthquake Research and
    Information, 157
chemical communication, 56
chloroquine, 44
Choyi, 41, 45, 98
CITES (Convention on the Interna-
    tional Trade in Endangered
    Species), 153
climate change, 56
cochlea, 57
Collar (matriarch elephant), 2, 8–9, 60
collars, see radio and satellite collars

communication:
    types of, 56
    see also seismic communication
community game guards, 32, 100–105,
    152–53
community resource monitor (CRM)
    program, 100, 102, 105–7, 133
Congo basin, 81
conservancies, 26
Convention on the International Trade
    in Endangered Species (CITES),
    153
corn, 40, 47, 50–51, 55, 84, 97
cortisol levels, 203
Costner, Kevin, 6
cow elephants:
    allomothering and, 63
    birth control experiments per-
        formed on, 83–84
    birthing mothers protected by, 17
    dominance among, 168
    gestation and nursing period of, 113
    greeting ceremonies among, 204–6
    outside author's window, 76–77
    teats of, 15
    see also estrus; matriarchs
crocodiles, 6, 23, 30–31, 87, 89, 112
crop-raiding elephants, 3, 24, 46, 55,
    84, 221
    author's work on, 44, 51–53,
        58–61, 127, 148
    "Dirty Seven" herd, 47–50
    electric fencing for, 51–53, 80,
        90–91, 105, 129, 132, 148,
        155–56
    family groups of, 61
    means of protection from, 40, 43,
        50–52
    murder by farmers and, 222
    playback of elephants' alarm calls
        and, 59–61
    three choices for, 155–56
    trip alarms for, 51, 53, 58–59
Curie, Marie, 171

Dances with Wolves, 6
Daniel the Star Gazer (bull elephant),
    221, 222
Dave (author's field assistant), 190, 191
David the Lion Slayer (bull elephant),
    211, 215
Davis, University of California at, 120
Dead Pan, 148
Democratic Turnhalle Alliance (DTA),
    45

elephants, (cont'd)
  surprises disliked by, 73, 77, 113
  toenails of, see toenails
  tree species eaten by, 81, 86
  Triangle population of, 22
  types of communication used by, 56
  vocalizations by, see vocalizations
    (elephant)
  water-storing pouches of, 110
  weight of, 48
  see also bull elephants; cow ele-
    phants; crop-raiding elephants;
    matriarchs; young bull elephants
elephant seals, 6
Elephus (Asian elephants), 82, 124,
  185
Emerson, Ralph Waldo, 1
Erin (cow elephant), 188
erioloba trees, 21, 74, 87, 90
Ernest (game guard), 45–47
estrogen implants, 83
estrus:
  bulls' searching for cows in, 8, 200,
    207, 208, 209–10
  cows protected by bulls during, 113
  duration and frequency of, 113
  mating in, 191
estrus calls, 68
  author's experimental playbacks of,
    187–89
Etosha Ecological Institute, 61, 187
Etosha National Park, 1–2, 3, 20, 27,
    39, 55, 75, 84, 94, 111, 133,
    147, 172, 186
  counting of elephants in, 24–25
  cull of elephants in, 157
  see also Gobaub tower; Mushara
    (watherhole); Olifantsbad
    (waterhole)
euphorbia, 51
European Union, 25, 28, 127

false apricots, 98
family groups:
  age composition of, 196
  and author's vocalization playback
    experiments, 59–61, 67–69,
    189–91, 195–208
  bond units made up of, 14
  crop damage by, 61
  dominance hierarchy and, 67,
    197–99
  at rest, 73
  reuniting of, 88
  spacing and, 179, 189

waterholes shared by, 2, 8–9, 62,
    63, 67, 210
  young bulls forced out of, 16
  see also matriarchs
farmers, 24, 40
  government elephant program and,
    45–46
  government employee killed by, 34
  initial resistance to author's crop-
    protection program of, 52–53
  modernization and, 51, 155
  wet-season huts of, 50
  wildlife ownership and, 25–26
  see also crop-raiding elephants
fat, acoustic, 48–49, 184
fetal elephants, 84
flamingos, 206
flies:
  mopane, 91
  tsetse, 86, 113–14
foot and mouth disease, 127
footfall detection, long-distance, 126
frequency modulation, 58
fur, 57

Gakulu (bull elephant), 208
Garth (IRDNC official), 100–104,
    149
gemsbok (oryx), 17, 19, 202–3
Geode, 188, 190
Geological Survey, U.S. (USGS),
    165–66, 181
geophones, 20, 124–25, 174, 177,
    181–82, 187–88
German colonialism, 21
gestation period, 113
Ghanian proverb, 97
giraffes, 200
global positioning systems (GPS), 51,
    173, 176
Gobaub tower, 67–69
gold, 123
Golden Highway, 90
  description of, 85
  fatal incident involving author on,
    132–44
  fatalities along, 129
  paving of, 172
  shooting of tourists along, 172
  wet-season driving conditions on,
    23, 85, 132–33, 137
golden moles, 6, 82
Golden Triangle, 109, 112
Gondwanaland, 147
grain, 46, 50

Miss Ellie (cow elephant), 128
molecular biology, 172
mopane flies, 91
"Mother of All Elephants," 97–98
Mozambique, 84
Mtombo (Bushman), 80
Mudella, 42
Mudumu elephant herd, 53, 91
Mudumu National Park, 91, 109, 114
Munali, Beaven, 148
Mushara (waterhole):
    author's elephant studies conducted
        at, 1–20, 59–61, 62, 67, 69,
        77–78, 171–82, 186–94,
        195–208, 209–16, 217–24
    author's hide (bunker) at, 2, 3–4,
        172, 219–20
    author's leaving of, 222–24
    author's research tower at, 171,
        173, 176, 186–87, 209, 219–20
    "cafe" at, 186
    description and location of, 2
    elephant inspections of author's
        equipment and bunker at,
        13–14, 172
    family groups' sharing of, 2, 8–9
    lions at, 17–19, 176, 198–99,
        200–203, 222
    public barred from, 2
musth, 188–89, 193, 218
    bulls' searching for estrus cows
        while in, 8, 200, 207, 208,
        209–10
    dominant bulls and, 214, 218
    signs of, 7–8, 189, 203, 211–12,
        215
    suppression of, 219
musuhili (greeting), 42–43, 151

Nairobi, 29
namazu (catfish), 166–67
Namib Desert, 23, 82, 122, 147, 148
Namibia, 2, 21, 56, 116, 127, 221
    army of, 120
    colonial and political history of, 21,
        22
    independence of, 22, 26
    secession movement in, 128
national parks, 114
Native Americans, 6–7, 24, 121
New Jersey, 121
Ngoma, 128
night apes (bush babies), 131
No Tusker (matriarch elephant), 63, 64

Nova floodplain, 109
nursing period, 113

Oakland Zoo, 126
    author's seismic communication
        experiments conducted at,
        157–69
Okaukuejo, 16, 66
Okavango delta, 21, 24, 112, 127, 147
Olifantsbad (waterhole):
    author's elephant studies conducted
        at, 61–66, 67
    bull elephant's inspection of
        author's recording equipment at,
        63–64
    lions at, 62–63, 65–66
Omega, 136–44
Oonts (Kipling), 209
"Operation Jumping Man," 126
Orville (bull elephant), 214
oryx (gemsbok), 17, 19, 202–3
Osh (bull elephant), 158–59, 164,
    167–68, 189
Owambo people, 2

*Parable of the Blind Men and the Ele-
    phant,* 183
pharyngeal pouch, 110
Piggery, 150
planthoppers, 5
plants, elephants' ecological relation-
    ships with, 81, 87
Pleistocene era, 82
Pliny the Elder (Gaius Plinius Secun-
    dus), 71
poaching, 3, 30, 31, 111, 115, 120,
    130, 141, 149–50, 152–53
Poole, Joyce, 188–89
Popa Falls, 114–15
post-traumatic stress disorder (PTSD),
    222
pressure receptors, 7
preventive medicine, 115
primates:
    pressure receptors of, 7
    vibration-detecting cells of, 184–85
Prince Charles (bull elephant), 204
protein immunization, 83–84
putrefaction bacteria, 92

quinine, 45

radio and satellite collars, 3, 8, 83, 85,
    109–17, 127, 205

ABOUT THE AUTHOR

Caitlin O'Connell is currently a research associate at Stanford University in the Department of Otolaryngology, Head and Neck Surgery (OHNS). After earning her undergraduate degree in biology at Fairfield University, she obtained her master's degree in ecology, evolution, and conservation biology at the University of Hawaii–Manoa before completing her Ph.D. in ecology at the University of California–Davis. Her discoveries published in academic journals have always been reported in the popular news media, including *National Geographic*, *Scientific American* (France), *Natural History*, *Discover*, *New Scientist*, *Science News*, and *The Economist*. O'Connell has been exploring the capacity of elephants to detect seismic cues for more than fifteen years, and her studies have contributed to ideas for potential therapies for hearing-impaired people using vibration treatment. She recently cofounded and directs the nonprofit organization, Utopia Scientific (www.utopiascientific.org), an organization dedicated to science education, public health, and conservation, as well as the production company Triple Helix Productions, with a mission of providing quality science content in media.